Monitoring Neuronal Activity

The Practical Approach Series

SERIES EDITORS

D. RICKWOOD

Department of Biology, University of Essex
Wivenhoe Park, Colchester, Essex CO4 3SQ, UK

B. D. HAMES

Department of Biochemistry and Molecular Biology,
University of Leeds, Leeds LS2 9JT, UK

Affinity Chromatography
Anaerobic Microbiology
Animal Cell Culture (2nd edition)
Animal Virus Pathogenesis
Antibodies I and II
Biochemical Toxicology
Biological Membranes
Biomechanics—Materials
Biomechanics—Structures and Systems
Biosensors
Carbohydrate Analysis
Cell Growth and Division
Cellular Calcium
Cellular Neurobiology
Centrifugation (2nd Edition)
Clinical Immunology
Computers in Microbiology
Crystallization of Proteins and
 Nucleic Acids
Cytokines
The Cytoskeleton
Diagnostic Molecular Pathology
 I and II
Directed Mutagenesis
DNA Cloning I, II, and III
Drosophila
Electron Microscopy in Biology
Electron Microscopy in Molecular
 Biology
Enzyme Assays
Essential Molecular Biology I and
 II
Fermentation
Flow Cytometry
Gel Electrophoresis of Nucleic Acids
 (2nd Edition)
Gel Electrophoresis of Proteins
 (2nd Edition)
Genome Analysis
HPLC of Macromolecules
HPLC of Small Molecules
Human Cytogenetics I and II
 (2nd edition)
Human Genetic Diseases
Immobilised Cells and Enzymes
Iodinated Density Gradient Media
Light Microscopy in Biology
Lipid Modification of Proteins
Liposomes
Lymphocytes
Lymphokines and Interferons
Mammalian Cell Biotechnology
Mammalian Development
Medical Bacteriology
Medical Mycology
Microcomputers in Biochemistry
Microcomputers in Biology
Microcomputers in Physiology

Monitoring
Neuronal Activity
A Practical Approach

Edited by
J. A. STAMFORD
Anaesthetics Unit, London Hospital Medical College, London

—at—
OXFORD UNIVERSITY PRESS
Oxford New York Tokyo

Oxford University Press
Walton Street, Oxford OX2 6DP

Oxford is a trade mark of Oxford University Press

Published in the United States
by Oxford University Press, New York

A catalogue record for this book is available from the British Library

Library of Congress Cataloging in Publication Data
Monitoring neuronal activity: a practical approach/edited by J. A.
Stamford.
p.   cm.—(Practical approach series)
1. Neurophysiology—Methodology.   2. Neurons.   I. Stamford, J. A.
II. Series.
[DNLM: 1. Electrophysiology. 2. Neurons. 3. Neuroregulators–
physiology. WL 102.5 M744]
QP356.M569 1991 591.1'88—dc20 91–20871
ISBN 0 19 963244 8 (h/b)
ISBN 0 19 963243 X (p/b)

Typeset by Dobbie Typesetting Limited, Tavistock, Devon
Printed by Information Press Ltd, Oxford, England

Preface

Although neurons generate electrical impulses (action potentials), the events of neurotransmission also involve ionic fluxes, changes in cell metabolism and neurotransmitter release. Nevertheless, when most neuroscientists think about neuronal activity, it is usually the *electrical* activity of nerve cells that comes to mind, in particular, the 'firing rate' of the cells. However, action potentials are only one facet of a neuron's *activity* and, by concentrating solely on this aspect, only a partial view is obtained. One ignores an enormous range of other chemical, metabolic, and ionic aspects of nerve cell activity. This book aims to provide a broad-based discussion of the many different ways neuronal activity may be assessed.

There are essentially four main approaches to measuring the activity of nerve tissue: electrical, neurochemical, metabolic, and ionic. Just as diverse are the levels of resolution of the various methods, from entire groups of cells down to individual ion channels.

The first part of the book (Chapters 1–4) discusses the electrical activity of neurons, taking as its starting point a chapter on extracellular action potential monitoring. Millar covers the manufacture of the various types of extracellular microelectrodes and gives up-to-date information on data processing. Following on in Chapter 2, Silinsky discusses the practical and theoretical aspects of intracellular recording, with particular emphasis on voltage clamp methodology. Voltage clamp resurfaces from a different angle in Chapter 3 where Standen and Stanfield extend the method to the single ion channel level in a practical discourse on patch clamping. Chapter 4 by Haglund and Blasdel offers a fascinating new slant on the electrical events of neurotransmission in which the activity of cortical neurons is directly visualized using dyes sensitive to changes in the membrane potential.

A correlate of cell activity characteristic of neurons is the release of neurotransmitters, the chemicals with which they communicate. The second part of the book (Chapters 5–7) covers three complementary methods of measuring transmitter release. Chapter 5 discusses *in vivo* voltammetric methods to monitor local extracellular transmitter levels, with practical details on two of the most widely used. Another high sensitivity method of increasing popularity is intracerebral dialysis in which transmitters are collected by tiny perfused microprobes. In Chapter 6, Sharp and Zetterström cover the manufacture, use, and application of dialysis probes. A further, fundamentally different, method for measurement of neuropeptide release is given in Chapter 7 by Duggan, in which microprobes are coated with antibodies to specific peptides, thus enabling selective measurement of individual peptides *in vivo*.

vii

Cell metabolism is a third means by which the activity of neurons may be assessed. The uptake of glucose by nerve cells provides a direct index of their metabolic workload and forms the basis of the 2-deoxyglucose radiographic method discussed by Kennedy, Smith, and Sokoloff in Chapter 8.

Action potential generation, neurotransmitter release, and the maintenance of resting membrane potential involves the flux of ions between intra- and extracellular compartments. Measurement of ion concentration thus provides the final facet of nerve cell activity and is discussed in the last two chapters. In Chapter 9, Duchen covers the new fluorescence technique in which fluorophores specific for individual ions are introduced into nerve cells, and thus allow one to monitor intracellular ion concentration. The chapter provides details on all aspects of the instrumentation. In the final chapter of the book Syková discusses the preparation and use of ion-selective microelectrodes, primarily for the measurement of extracellular ions, providing an ideal counterpoint to the previous chapter.

It is hoped that this book will help to make the reader aware of the many facets of nerve cell activity and the ways in which they may be studied, from the opening and closing of a single ion channel through to the unified activity of whole brain areas, from the flux of ions through the cell membrane to the release of neurotransmitters. I should particularly like to thank the contributors who, each in their own way, have described their method with clarity and in detail. I hope that they convey as much of their enthusiasm to the reader as they have to the editor.

London   J. A. STAMFORD
April 1991

Contents

3. Patch clamp methods for single channel and whole cell recording   59

N. B. Standen and P. R. Stanfield

4. Video imaging of neuronal activity   85

Michael M. Haglund and Gary G. Blasdel

5. *In vivo* voltammetric methods for monitoring monoamine release and metabolism

J. A. Stamford, F. Crespi, and C. A. Marsden

6. *In vivo* measurement of monoamine neurotransmitter release using brain microdialysis

Trevor Sharp and Tyra Zetterström

7. Antibody microprobes   181

A. W. Duggan

Contents

Contributors

GARY G. BLASDEL
Department of Neurobiology, Harvard Medical School, Boston, MA, USA.

F. CRESPI
Department of Physiology and Pharmacology, University Hospital and Medical School, Clifton Boulevard, Nottingham NG7 2UH, UK.

MICHAEL R. DUCHEN
Department of Physiology, University College London, Gower Street, London WC1E 6BT, UK.

A. W. DUGGAN
Department of Preclinical Veterinary Sciences, Royal School of Veterinary Studies, Summerhall, Edinburgh EH9 1QH, UK.

MICHAEL M. HAGLUND
Department of Neurological Surgery, University of Washington, Seattle, WA, USA.

CHARLES KENNEDY
Laboratory of Cerebral Metabolism, National Institute of Mental Health, US Department of Health and Human Services, Public Health Service, Bethesda, MD 20892, USA.

C. A. MARSDEN
Department of Physiology and Pharmacology, University Hospital and Medical School, Clifton Boulevard, Nottingham NG7 2UH, UK.

J. MILLAR
Department of Physiology, Faculty of Basic Medical Sciences, Queen Mary and Westfield College, Mile End Road, London E1 4NS, UK.

TREVOR SHARP
MRC Unit of Clinical Pharmacology, Radcliffe Infirmary, Woodstock Road, Oxford OX2 6HE, UK.

E. M. SILINSKY
Department of Pharmacology, Northwestern University Medical School, 303 East Chicago Avenue, Chicago, IL 60611-3008, USA.

CAROLYN BEEBE SMITH
Laboratory of Cerebral Metabolism, National Institute of Mental Health, US Department of Health and Human Services, Public Health Service, Bethesda, MD 20892, USA.

Contributors

LOUIS SOKOLOFF
NIMH, 9000 Rockville Pike, Building 36, Room 1A-05, Bethesda, MD 20892, USA.

J. A. STAMFORD
Anaesthetics Unit, London Hospital Medical College, Whitechapel, London E1 1BB, UK.

N. B. STANDEN
Ion Channel Group, Department of Physiology, University of Leicester, PO Box 138, Leicester LE1 9HN, UK.

P. R. STANFIELD
Ion Channel Group, Department of Physiology, University of Leicester, PO Box 138, Leicester LE1 9HN, UK.

E. SYKOVÁ
Czechoslovak Academy of Sciences, Institute of Physiological Regulations, Bulovka, Pav 11, 180 85 Prague 8, Czechoslovakia.

TYRA ZETTERSTRÖM
Oxford University Beecham Centre for Applied Neuropsychobiology, Radcliffe Infirmary, Woodstock Road, Oxford OX2 6HE, UK.

Abbreviations

The list of abbreviations given below is the sum of those found throughout the ten chapters. It does *not* include:

A: Non-specific abbreviations such as the use of 'k' to define any constant in any context.

B: Abbreviations that are confined to figures and defined within the figure legend, and not used anywhere else in the text.

C: Abbreviations where individual authors may have used particular abbreviations in a manner idiosyncratic to their own chapters.

The list, therefore, defines abbreviations that have the same (i.e. universal, within the context of the book) meaning wherever they appear.

A	ampere
Å	angstrom
ACh	acetylcholine
A/D	analog to digital
Ag/AgCl	silver–silver chloride
APTES	gamma-aminopropyltriethoxysilane
BSS	balanced salt solution
C	capacitance
CCD	charge-coupled devices
CT	computerized tomography
C_t	electrode tip capacitance
C_w	wire capacitance
cm	centimetre
Cm/C_m	membrane capacitance
CNS	central nervous system
CSF	cerebrospinal fluid
DA	dopamine
D/A	digital to analog
dB	decibel
DCC	discontinuous current clamp
DG	2-deoxyglucose
DG-1-P	2-deoxyglucose-1-phosphate
DG-6-P	2-deoxyglucose-6-phosphate
DOPAC	3,4-dihydroxyphenylacetic acid
DPV	differential pulse voltammetry
ECD	electrochemical detection

ECG	electrocardiogram
ECS	extracellular space
EDTA	ethylenediamine tetraacetic acid
EEG	electroencephalogram
EGTA	ethylene glycol tetraacetic acid
E_{ljp}	liquid junction potential
$E_{\Delta ljp}$	difference in liquid junction potential
EMF	electromotive force
esp	excitatory synaptic potential
E_{tip}	electrode tip potential
$E_{\Delta tip}$	difference in tip potential
F-6-P	fructose-6-phosphate
FCV	fast cyclic voltammetry
FET	field effect transistor
Gm	membrane conductance
g	gram
GΩ	gigaohm
G-6-P	glucose-6-phosphate
G-6-Pase	glucose-6-phosphatase
GTP	guanine triphosphate
5-HIAA	5-hydroxyindoleacetic acid
HPLC	high performance liquid chromatography
HRP	horseradish peroxidase
5-HT	5-hydroxytryptamine
HVA	homovanillic acid
Hz	hertz
I/i	current
ICS	intracellular space
I_m	membrane current flow
i.m.	intramuscular(ly)
i.p.	intraperitoneal(ly)
ISM	ion-selective microelectrode
i.v.	intravenous(ly)
kΩ	kilo-ohm
K_D	dissociation constant
kg	kilogram
l	litre
m	metre
mA	milliampere
MAO	monoamine oxidase
MEM	modified Eagle's medium
mg	milligram
MHPG	3-methoxy-4-hydroxyphenylglycol
min	minute

ml	millilitre
mm	millimetre
mΩ	milliohm
MΩ	megohm
MRI	magnetic resonance imaging
ms	millisecond
mV	millivolt
nA	nanoampere
NA	noradrenaline
nC	nanocoulomb
nm	nanometre
nF	nanofarad
op. amp.	operational amplifier
pA	picoampere
PBS	phosphate-buffered saline
PCM	pulse code modulator
PET	positron emission tomography
pF	picofarad
PMT	photomultiplier tube
PVC	polyvinyl chloride
R	resistance
RC	resistor–capacitor
ref.	reference
REM	rapid eye movement
R_{in}	input resistance
R_m	membrane resistance
R_{me}	microelectrode resistance
RMS	root mean square
R_{tip}	electrode tip resistance
s	second
SEVC	single electrode voltage clamp
SIT	silicon-intensified target
SPECT	single photon emission computerized tomography
SST	somatostatin
TDDA	tridodecylamine
TEA	tetraethyl ammonium
TEVC	two electrode voltage clamp
TMA	tetramethyl ammonium
TTL	transistor–transistor logic
UDP-DG	uridine diphospho deoxyglucose
V	volt
V_c	command voltage
V_{eq}	equilibrium potential
V_m	membrane potential

Ω	ohm
μA	microampere
μl	microlitre
μm	micrometre
μs	microsecond
μV	microvolt

1

Extracellular single and multiple unit recording with microelectrodes

J. MILLAR

1. Introduction: the origins of extracellular spike potentials

Action potentials across the membranes of neurons generate electric current flow in the tissue around the neurons. These currents in turn generate voltages in the tissue, and these voltages can be detected as extracellular 'spikes'. Extracellular spikes from mammalian neurons in the central nervous system have a duration of between 0.2 and 20 ms, depending on the type of neuron and the bandwidth of the spike recording system. Amplitudes vary from the noise level of the recording electrode (see below) which may be as little as 2 μV RMS (root mean square value) up to several millivolts. The great advantage of extracellular recording for the neurophysiologist is that you can record the activity of neurons without having to impale and thereby damage them. Impalement may be inadvisable because it damages the cell, it may be extremely difficult because the cell is small, or it may be unstable because the tissue is moving due to mechanical pulsations from blood pressure or respiratory fluctuations. For these reasons most neuronal recording in anaesthetized animals, as opposed to brain slices or cell cultures, is done with extracellular recording.

The great disadvantage of extracellular recording is that it is all too easy to record from two or more units at the same time; this is technically 'multi-unit' recording. This, of itself, is not a problem if you can discriminate between the spikes from the two (or more cells). This, however, may be difficult. It is rather like having a microphone between two people talking together in a noisy environment. Which voice is active at any one moment may be determined simply by the loudness of the voice; however, if the voices are of a similar loudness (amplitude) other characteristics such as the pitch of the voice and perhaps the speed of speaking have to be taken into account.

Similarly, the presence of a spike at any given moment does not guarantee that it is generated from the same cell that produced a spike 40 ms ago; we must discriminate between spikes on the basis of spike amplitude, shape, and possibly firing pattern, and only accept that the same cell is being recorded if we can

1

establish a fixed spike shape or pattern sufficiently different from anything else that is detectable at that particular recording site. Because extracellular spikes may be only just above the noise level, a key parameter in extracellular electrodes is the electrode noise. The lowest noise extracellular electrodes currently in use are the recently-introduced silver-plated carbon-fibre electrodes. Platinum-plated tungsten electrodes can also be very low-noise and have been used successfully for a number of years. Glass micropipettes are the easiest extracellular electrodes to manufacture but are considerably more noisy than the two solid-conductor types just mentioned. Details of how to manufacture these three electrode types are given below. Several other types of extracellular electrodes are used for special purposes (for example microwire for chronic recordings); construction details for these electrodes are given in the references cited.

2. Tungsten microelectrodes

2.1 Background

Tungsten microelectrodes are the 'classic' extracellular electrode. They were first developed in the 1950s and have a long history of successful use in analysis of neuronal function. The Nobel prize-winning work of Hubel and Wiesel on the visual system was based on work done with tungsten microelectrodes.

2.2 Advantages

Microelectrodes obviously need to be of small diameter, at least near the tip, to minimize the mechanical damage that they will inflict as they are pushed through the neuronal tissue. The electrodes also need to be as stiff as possible so that they do not bend over when pushed in. A flexible electrode can 'drift' considerable distances sideways from its desired trajectory, which means that stereotaxic placement is difficult. Tungsten is a metal which is easily available and cheap in wire form (it is used to make electric light filaments), and is very stiff and strong.

2.3 Disadvantages

The main disadvantages of tungsten microelectrodes are

- They can be electrically noisy, unless plated with platinum black.
- Despite claims to the contrary in some research papers, they can be difficult to make.
- The electrochemical qualities of tungsten are such that these electrodes are un-suitable for chronic recordings unless the surface has been coated with platinum.

2.4 Availability

Tungsten microelectrodes come in two main forms; varnish-insulated and glass-insulated. Varnished tungsten electrodes were described by D. H. Hubel in 1957 (1), but difficulties in producing a constant tip recording area led researchers to develop the glass-insulated electrode which can now be regarded as the standard form. The manufacture of glass-insulated tungsten electrodes involves a certain investment in specialized apparatus but this is well worth it in the long run. An additional advantage is that the electrodes may also be used for microstimulation.

2.5 Manufacture of glass-insulated tungsten electrodes

2.5.1 Equipment
You will need the following equipment:

- old microscope with ×10 eyepieces and ×10 objective
- tungsten wire etching apparatus
- vertical type microelectrode puller
- two micromanipulators to move electrodes and platinum hooks into field of microscope
- electrode plating apparatus

An old microscope is recommended for observation of the second etch of the tungsten wire and for the plating processes. Etch and plating solutions invariably splash around and the microscope stage will eventually get a bit corroded, so an old model (look for ones being thrown out by the anatomy department!) is best. It should have a total magnification of 100 or 200, for example ×10 eyepiece and ×10 and ×20 objectives. Additional (incident) illumination is very useful and can be produced by a fibre-optic light source angled on to the stage from the side.

The etching apparatus has to dip the electrodes repetitively into a bath of etching fluid until they achieve the correct taper. For pilot work, or manufacture of a small number of electrodes, it is perfectly possible to do this by hand. For regular manufacture either of two types of etcher can be used. Both types need a holder for the electrodes during etching. This is a round brass spindle, 2 cm long by 2 cm diameter (*Figure 1*). It is drilled with a socket on one face to take a male Luer pattern tapered shaft. The electrodes are fixed to the circumference of the spindle by micropore tape. In one form of etcher (2) the spindle is mounted on a rotating shaft held at an angle so that the electrodes dip into the etch bath as the shaft rotates. This arrangement, however, can produce electrodes with oval cross-sections. A more reliable form of etch is obtained by using a vertical dipping process. The spindle is plugged into a male Luer fitting which is suspended from a strong thread which passes via a guide eye up to a lever attached to a rotating disk (*Figure 2*). The disk is driven at 5 r.p.m. by a small motor. Two small projecting lugs are attached to the circumference of the disk. These operate

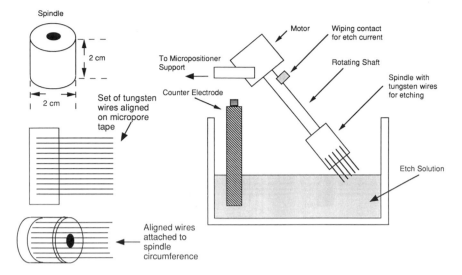

Figure 1. Apparatus for etching tungsten electrodes using angled dip. The tungsten wires are fixed to the circumference of a metal spindle fixed to the end of a rotating angled shaft. The shaft is then lowered until the wires dip into the etch solution for part of the rotation cycle. Electrical contact with the shaft is made by a wiping contact from a phosphor–bronze spring.

a microswitch to turn the etching current on and off via a relay. The lever moves the spindle up and down above the etch bath so that the electrodes are moved up and down in the fluid. The electrode wires are fixed around the circumference of the spindle, pointing down into the bath. To make a jig for mounting the wires, cut the socket ends off 10 or 15 large (250 µm internal diameter) hypodermic needles leaving 1 cm lengths at the tip end. Align the cut ends and glue these tubes in parallel at 5 mm spacings on a flat piece of wood or Perspex.

The etching fluid is a solution of potassium nitrite, 150 g/100 ml. It is contained in a shallow glass beaker or trough of 100 ml capacity. A carbon rod (or better, a set of carbon rods wired in parallel) dipping into the fluid makes up the counter electrode for etching. To carry out the etch, an AC voltage source of about 6 V at several amperes (A) is required. This can be obtained most simply by using a 'Variac' mains transformer to get a variable AC voltage of 0–250 V, and then stepping this down with a small isolating transformer with a nominal 12 V (5 A) output. The voltage is connected to the spindle and carbon rods by multistrand flexible cables ending in crocodile clips.

For the second etch a platinum wire (100 µm) is fixed between two small posts attached to the microscope stage or a rod fixed to a micromanipulator. The central part of the wire is formed into a small loop about 1 mm in diameter. This can hold a drop of etching fluid by surface tension. The loop can be moved backwards and forwards in the view of the microscope. A second manipulator

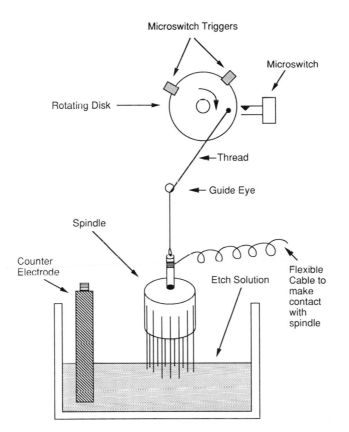

Figure 2. Apparatus for etching tungsten electrodes using vertical dip. The tungsten wires are fixed to the circumference of a spindle suspended from a strong thread attached via a guide eye to the edge of a vertical disk. They are moved up and down into the etching solution by rotation of the disk. A microswitch switches the etch voltage on and off so that etching only occurs at a predetermined part of the movement.

holds the spindle carrying the wires from the first etch. The same etching voltage can be used for the second etch, but a more dilute fluid is used (see below).

To coat the etched wires with glass a heavy-duty electrode puller suitable for the manufacture of iontophoresis (multi-barrel) electrodes is best. We use the Narishige type PE-2. A friction collar made of foam rubber around the base of the shaft where it enters the solenoid can be used to slow down the rate of fall of the shaft if necessary. A clamp attached to one of the main supports can also be used as a brake to stop the pull after a brief initial elongation of the glass.

For removing the glass tip of the electrode a second platinum loop is necessary, containing a small blob of solder glass (for example Schott type 8465) (*Figure 3*). If solder glass is unobtainable a blob of borax may be used but does not

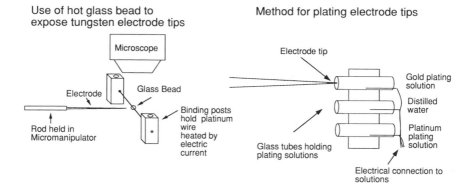

Figure 3. Apparatus for removing glass from tip of tungsten electrodes and plating. The freshly-pulled electrode with a glass whisker protruding beyond the metal tip is fixed in a micromanipulator and moved slowly into a molten glass bead. The bead is produced by electrically heating a small blob of solder glass on a nichrome or platinum wire. The glass is then allowed to cool and a crack will occur at the edge of the bead where the electrode meets it. The bare metal tip can then be withdrawn.

always work as well. The bead is heated until it melts but does not drip off the loop. This is done by applying an AC voltage of 0–5 V. (**Caution!!** too high a voltage will vaporize the wire and glass very rapidly.)

The plating apparatus requires a fine tube to hold the plating solution. This can be made from a piece of 2 mm diameter electrode glass pulled to make a glass microelectrode and then broken back to a tip diameter of about 100–300 μm. The pipette is filled with plating solution and a platinum wire pushed into the blunt end to make electrical contact. For plating, a DC current of between about 1 and 100 nA is needed. A useful source of current is an old iontophoresis machine, otherwise a 0–5 V variable DC voltage source in series with 10 MΩ resistor can be used. The tip of the plating tube is held on a micromanipulator so that it can be moved into the view of the microscope. Plating solutions for gold and platinum black are needed. It is useful to have a set of tubes containing the different plating solutions fixed together on a small rod attached to a micromanipulator so they can be moved under the microscope. Each tube contains a platinum wire, the other end of which is attached to a common terminal (*Figure 3*).

A suitable gold plating solution can be obtained commercially (Johnson Matthey). However, the platinum plating needs to produce a platinum black finish to the surface, and most commercial platinum plating solutions produce a bright platinum finish. Because of this it is recommended that you make your own platinum plating solution. The gold plating solution is also easily made in the laboratory, and will be much cheaper than a commercial solution (*Protocol 1*).

Protocol 1. Preparation of electrode plating solutions for tungsten electrodes

Gold plating

1. Dissolve 1.0 g hydrogen tetrachloroaurate (chloroauric acid $H_2AuCl_4 \cdot xH_2O$) in 71 ml distilled water.

2. Add concentrated ammonia solution or bubble with ammonia until precipitation complete.

3. Filter and wash the precipitate in distilled water. Dissolve in 143 ml of 1.25% potassium cyanide (KCN). Boil until red litmus held over the solution stays red or only turns blue very slowly. The solution will keep indefinitely at room temperature in a stoppered bottle.

Platinum plating

1. Dissolve 1.0 g hydrogen hexachloroplatinate (chloroplatinic acid $H_2PtCl_6 \cdot 6H_2O$) in 50 ml distilled water.

2. Mix 1.125 ml conc. HCl with 200 ml distilled water. Take 100 ml of this HCl solution and add 0.625 g lead acetate. Take 10 ml of this and remix with another 90 ml of the HCl solution; this gives a mixture of lead acetate and lead chloride.

3. To make the stock 'Kohlrausch' solution add 2 parts of this lead acetate–lead chloride solution to 3 parts of the chloroplatinic acid solution (step 1).

4. The plating solution is 1 part Kohlrausch solution with 1 part 0.17% aqueous gelatine solution.

2.5.2 Materials

Straight lengths of tungsten wire can be obtained from Clark Electromedical (Pangbourne, UK). The best diameter wire for electrodes is 125 μm. The wires should be between 6 and 10 cm long. This diameter tungsten wire can be cut to length but should always be cut with large scissors, not wire cutters (the scissors cut with a shearing action instead of a crushing action). Long lengths of wire on a spool can be obtained from Lamp Metals Ltd. The cost of the Lamp Metals wire is much less, but it has to be straightened by heating under tension (passing an AC current can be used to heat the wire; a current of 2–4 A should be sufficient under a tension of 2 kg). Chemicals for making up the plating solutions can be obtained from Aldrich. Standard microelectrode borosilicate glass (in the UK from Clark Electromedical) with a 1.5 or 2.0 mm outside diameter can be used for coating. Alternatively, Jencons type H 15/10 glass tubing can be used. Glass with an internal filament (for ease of filling the pulled pipette with electrolyte) should not be used.

Protocol 2. Glass-coated tungsten electrode manufacture

1. Cut the straight tungsten wire (125 μm diameter) into a minimum of 6 cm lengths. For penetration of deep brain structures in larger animals longer lengths may be used.

2. Place a batch of 10–15 wires into the jig so that their ends are aligned. Place a piece of micropore tape across the wires, lift them out of the jig and fix them around the circumference of the etching spindle.

3. Etch the wires using either the arrangement shown in *Figure 1* or *Figure 2*. With the dipper, the lugs operating the microswitch should be adjusted so that the etch current comes on only during the upward movement of the electrodes in the bath.

4. Wash the etched wires and fix the spindle so that it can be rotated under the microscope to view each wire in turn. Place a drop of diluted etching fluid (50 g per 100 ml) on the platinum loop and move one electrode into the drop. Apply 1–2 V AC between the wires and etching fluid to produce a sharper taper over the terminal 50–100 μm of the wire. This removes any fine whiskers of tungsten left from the first etch. The electrode does not have to be moved during this process. (The meniscus of the fluid will move down the tip of the electrode during the etch). After etching the electrode may be given a coating of black metal oxide, which appears to improve the glass–metal bond when it is coated with glass. To do this pass the etched wire, including the tip, a few times through a Bunsen burner flame for about half a second.

5. Block the end of a tube of electrode glass, 2 mm outside diameter, with a small plug of wax or plasticine. Drop the sharpened wire, butt end first, into the glass 'blank'. Fix the bottom plugged end of the glass in the chuck of the electrode puller, and adjust the position of the chuck so that the heating element is 1 cm or so below the sharp end of the electrode. If necessary cut the glass tubing down. The puller must be adjusted to give the gentlest pull possible. Disable the solenoid and let the pull proceed by gravity alone. If necessary extra frictional braking can be achieved by a collar of foam rubber around the base of the puller shaft. In some cases a single pull will suffice to collapse the glass around the electrode to form the required surface layer. However, if the coating is too thick, or does not coat a sufficient length of the electrode, a double pull can be used. Fix a clamp to one of the uprights of the puller so that the chuck hits this clamp and stops the initial pull after about 5 mm elongation. Move the element down so that it can now heat the region of narrower elongated glass. Remove the clamp and start the second pull. With practice a suitable set of pull parameters will be obtained for any particular puller and from then on reproducible coatings should be obtained. A whisker of glass should protrude beyond the tip of the electrode after coating. If this is more than a few millimetres long cut if off with fine scissors.

6. Hold the coated electrode on a rod attached to a micromanipulator (a blob of modelling clay will do) so that the tip may be viewed under the microscope (*Figure 3*). Move the whisker up until it presses gently against the solder glass bead. Slowly increase the bead heater current until the bead softens. Push the electrode the required distance into the bead (say 20 μm) and turn off the current. It may be necessary to advance the electrode slightly as the bead cools so that the angle between the bead circumference and the electrode shaft is as sharp as possible. As the bead cools the glass should crack at this junction, the electrode can then be retracted leaving a 20 μm length of bare metal tip. If the process does not work on the first go reheat the bead and try again.

7. Either break back the glass stem of the electrode to expose the butt end of the tungsten wire or insert a small flexible wire dipped in conducting silver paint into the glass stem so as to make electrical contact with the tungsten (see *Protocol 3*, carbon-fibre electrodes). Bare wire electrodes are best held by crimping the butt end slightly and then pushing this end into the tip end of a hypodermic needle. The hypodermic needle can then be held on a micromanipulator on a rod with a Luer end fitting. (If the glass stem is not broken back this stem can be held in a conventional electrode holder).

8. Move the tube containing gold plating solution under the microscope and insert the tungsten electrode tip into the solution. Plate with about 250–500 nC, electrode negative (250 nA for 1–2 s or 50 nA for 5–10 s), or until the surface is just visibly yellow.

9. Wash the tip with water and move the tube with platinum plating solution into view. Plate with 300 nA (electrode negative) for about 1 minute, or until the tip is noticeably blackened.

10. Bare-end finished electrodes can be stored in slits cut in a foam rubber block, which is glued into the base of a suitable small box. Intact stem electrodes are best stored on individual microscope slides, attached by a blob of modelling clay.

Key references for tungsten electrodes; (2) and (3).

---

3. Carbon-fibre microelectrodes

3.1 Background

In the 1960s and 1970s the technique of microiontophoresis became very popular. This process involves the ejection of nA currents through one or more barrels of a multibarrel glass microelectrode so that ionized drugs in solution in the barrels can be ejected and applied locally to a single nerve cell or group of cells.

This drug application is combined with extracellular recording through one barrel. However, glass recording electrodes which are included in a set of iontophoretic barrels are notoriously electrically noisy. Because of this, many researchers have tried to combine a tungsten electrode with a set of iontophoresis barrels. The usual method has been to glue the two types of electrode together, under an operating or dissecting microscope; this is a difficult operation. Alternatively, one can try inserting a presharpened tungsten wire into the barrels, but again, this is not so easy as it sounds. Carbon-fibre electrodes were developed in the late 1970s as a means of combining the good extracellular recording qualities of tungsten-type electrodes together with iontophoretic barrels in one easy-to-assemble multibarrel assembly. Single carbon-fibre electrodes (i.e. without iontophoresis barrels) were found to have electrical properties similar to the best tungsten electrodes.

3.2 Advantages

Carbon fibres are available commercially in 'tow's of many thousands of fibres of 7–8 μm diameter for a very low price. They are good conductors and form a particularly low-noise interface with electrolyte solutions. Carbon-fibre electrodes are easier to make than tungsten electrodes and require less specialized apparatus. One useful characteristic of carbon-fibre electrodes is that they can also be used for voltammetric analysis of transmitters *in vivo*. For example using Fast Cyclic Voltammetry (FCV) (see Chapter 5), carbon-fibre electrodes can be used to measure the release of dopamine (DA) in the rat caudate following electrical stimulation of the DA-secreting nigrostriatal neurons. It is possible, using a time-share system, to alternate automatically between FCV analysis and unit recording, opening up the possibility of sampling the 'chemical activity' of neurons at the same time as their electrical activity (4). The technique for making voltammetric-recording carbon-fibre electrodes is the same as for unit recording electrodes, except that voltammetric electrodes should not be spark-etched (see Section 3.4), as this makes them electrochemically noisy.

3.3 Disadvantages

Carbon-fibre electrodes are not as stiff as tungsten electrodes because the carbon has a uniform diameter instead of a taper back from the tip. They are not as strong as tungsten and are more easily damaged; for example there is also a tendency for the glass insulation at the tip to either break off or become leaky, and this limits the number of times they can be reused. One analogy which may be useful is to think of tungsten electrodes as being rather like the old-style precision glass hypodermic syringes; they were very accurate, were reusable indefinitely, but were expensive and difficult (for the manufacturer) to make. Carbon-fibre electrodes are like modern plastic hypodermics; cheap, easy to make, but not meant to be reused.

3.4 Manufacture of carbon-fibre electrodes

You will need:

- a heavy-duty microelectrode puller (see above for tungsten electrodes)
- a draught-proof area of bench for inserting the fibres in the glass blanks
- tools for handling the fibres
- apparatus for cutting or etching the carbon tips

For multibarrel carbon-fibre electrodes a heavy-duty puller capable of pulling iontophoresis electrodes is necessary. For single carbon-fibre electrodes horizontal pullers used for glass micropipettes may be adequate. (The author has not tried to make electrodes on any but the vertical pullers made by Narishige or Harvard Instruments.)

Individual fibres have to be separated from the multifibre tow and inserted into a piece of glass electrode tubing. The best surface to do this on is a flat-topped light box as used by photographers to inspect slides. The box should have as large an area as possible, say 0.5 m × 0.5 m and be placed on a bench where there are no draughts. To handle the fibres you need a pair of no. 5 jewellers forceps which have small plastic collars over the two tips. The plastic collars can be made by stripping short (2–5 mm) lengths of insulation from small signal wires.

There are several ways of trimming the end of the carbon fibre to the correct tip length. The simplest is to cut it off with microscissors under a microscope. This is still mandatory for electrodes to be used for voltammetric analysis. However, the method that gives the maximum control of tip length for selective unit recording is spark etching. For this you need a variable-amplitude pulse generator which can produce 50–500 µs pulses at up to 100 per second. This must have sufficient output current capacity to drive a step-up transformer to give a final pulse voltage of 1000–1500 V.

Protocol 3. Manufacture of single carbon-fibre electrodes

1. Cut a hank of fibres about 15 cm long from the tow. Place the hank in a test-tube so that 2–5 cm of the bundle protrudes from the end. Fix the test-tube horizontally on the light box so that individual fibres may be withdrawn using the plastic-tipped forceps.

2. Put a piece of 2 mm diameter electrode glass tube (10 cm long) in a small test-tube held at an angle of a few degrees to the horizontal by a blob of plasticine. The electrode glass should protrude out of the end of the tube. Use a dropping (Pasteur) pipette to squirt a few millilitres of acetone into the test-tube. The acetone should immediately fill the glass tube by capillary action.

3. Splay out the end of the fibre bundle and pull a single fibre from the bundle with the forceps. Holding the fibre about 1 cm back from the end, push

Protocol 3. *Continued*

the end carefully into the acetone-filled electrode glass tube (*Figure 4*). If the level of illumination is correct most people can see single fibres with their naked eyes. For the hypermetropic researcher, a low-power dissecting microscope may be useful. Once the end of the fibre is in the glass tube, release the forceps and grip it again about 1 cm further back. By this means the fibre can be pushed little by little into the tube. When the fibre is fully in the tube, the protruding end may be cut off with fine dissecting scissors. The glass blank should now be removed from the test-tube and one end blotted carefully with tissue to remove the acetone. Once the acetone is removed, there is little further risk of the fibre falling out; the carbon fibre should adhere to the side of the glass by surface tension. Allow the acetone to evaporate for at least one hour, ideally overnight.

4. Pull the filled glass blank in the puller with the solenoid disabled so that the chuck drops by gravity alone. You should end up with the carbon fibre gripped tightly by the collapsed glass in one of the two pipettes formed, and pulled totally or nearly totally out of the other one. The great tensile strength of the carbon fibre prevents it breaking when the glass fractures to form two pipettes. Cut the protruding fibre back to a few millimetres beyond the end of the pipette with the longer length of fibre inside, and transfer this pipette to a micromanipulator so that the tip can be viewed under a microscope ($\times 100$ or $\times 200$ magnification).

5. Visually check that the glass forms a tight seal at the end of the electrode, and also that a sufficient length of carbon fibre protrudes back into the stem of the pipette to make a contact when the connecting wire is inserted.

6. Strip back 1–2 cm of insulation from a piece of miniature stranded signal wire and cut all but one strand off, so that a very fine tinned copper wire

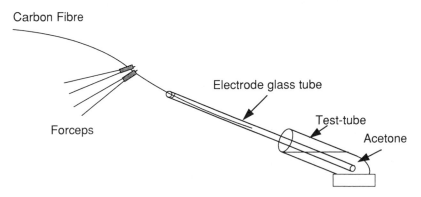

Figure 4. Method of insertion of carbon fibres into glass blanks. A small test-tube is held at a shallow angle and half-filled with acetone. The electrode blank is inserted and fills with acetone by capillary action. A carbon fibre is inserted using plastic-tipped jewellers (no. 5) forceps.

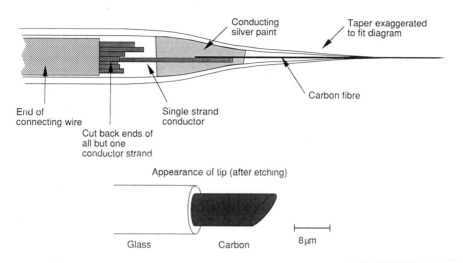

Figure 5. Connection of lead wire to carbon fibre electrodes. A fine plastic-coated multistrand signal wire is stripped of insulation for 1–2 cm and all but one conductor trimmed back to the insulation. This fine wire is dipped into silver-doped conducting glue and the wire rapidly inserted into the butt end of the electrode. The fine wire should form a strong joint with the carbon fibre.

is left. Dip this into a solution of conducting silver glue and immediately push it into the stem end of the electrode so that the wire makes contact with the stem end of the carbon fibre (*Figure 5*). The wire should be small enough so that the insulated part can also be pushed some way into the glass. Anchor and seal the end, where the insulation emerges, with a drop of cyanoacrylate glue.

7. If the electrodes are to be used for voltammetry the carbon fibre protruding beyond the end of the glass must be cut back to a length of 20–50 μm with microscissors. A pair of the smallest available scissors (iridectomy scissors are suitable, or those used for microvascular surgery) can be used hand-held under the microscope to trim the carbon fibre to the required length. (This procedure does, however, require an unusual lack of physiological tremor in the operator to avoid a high failure rate!). Alternatively, one arm of the scissors can be mounted on a micromanipulator and the blades closed in a controlled manner by a screw fixed to push down on the other arm. The scissors produce a surprisingly clean 'snap' cut of the fibre, probably for the same reason that tungsten wires are much better cut with scissors; the shearing action of scissors is better than the crush action of wire cutters.

7a. If the electrode is not needed for voltammetry it can be spark-etched. Bring the tip of the electrode with the fibre sticking out up to the end of a tapered tungsten wire (see above in section on tungsten electrodes) so that the two conductors just touch 100 μm beyond the glass insulation. Apply high voltage pulses between the two conductors until a small spark is just visible under

Protocol 3. *Continued*

the microscope. (Use the lead wire to make contact with the carbon fibre.) A correct sized spark will gradually eat through the carbon fibre until it is severed at the contact point. For extracellular recording, etch to a tip length of 10–30 μm.

8. For extra low noise recording (again, this process is not suitable for voltammetric electrodes) the carbon tip can be coated with silver/silver chloride. To do this, suspend a drop of dilute silver nitrate (0.05 M) from a platinum loop and insert the carbon tip into the drop. Pass 10 μA RMS current for 1–5 s between the carbon and solution until the carbon tip has a 'fuzzy' appearance. This is due to the presence of finely divided silver on the tip. Wash the tip and leave it to soak in a solution of 0.9% saline for a few minutes. In the presence of chloride ions some of the silver reacts to form silver chloride which thus forms a silver/silver chloride interface with the solution. This process reduces the interface impedance and hence noise at low frequencies.

9. Carbon-fibre electrodes can be stored by pressing the stem on to a blob of modelling clay stuck on a microscope slide. The slides can then be stored in a slide box; electrodes may be tested for electrical noise while still fixed on to their carrier slides.

The manufacture of carbon-fibre electrodes for iontophoresis, as well as unit recording, follows the steps described above for single electrodes, except that 'theta' or 'quadrant' microelectrode glass blanks should be used. The fibre is pushed into one barrel of the quadrant, or theta, glass blank using acetone as normal. The electrode is then pulled as for the single electrodes. Multibarrel electrodes have a tendency to bend at the end due to the uneven stresses in the glass, and to alleviate this a two-stage pull (see 'tungsten electrodes') may be necessary. The iontophoresis barrels are limited to three in this process, but the success rate for these electrodes is far higher than that obtained when separate glass blanks are glued together and then pulled with a twist pull. Only a confirmed masochist would prefer to make electrodes from separate glass blanks now that quadrant glass is easily available. Key references for carbon-fibre electrodes; (5–7).

4. Micropipette extracellular electrodes

4.1 Background

Micropipettes are the simplest form of extracellular microelectrode and have been in use for many years.

4.2 Advantages

Simplicity of manufacture, low cost, and reproducibility of performance are the main advantages of these electrodes.

4.3 Disadvantages

Electrolyte-filled pipettes are considerably more noisy than either tungsten or carbon-fibre electrodes. On the other hand, they can often detect larger extracellular spikes than the solid conductor electrodes by capacitance coupling with the recorded cell. The spikes detected by pipettes are more easily lost by small mechanical movement than those detected by solid conductor electrodes. The electrodes are unreliable for long periods of recording (more than an hour or so) as the electrolyte in the tip can diffuse out and change the recording qualities. Also the tip can 'block', that is go abruptly to a very high impedance. Glass pipettes are less stiff than carbon or tungsten electrodes, and so are not suitable for stereotaxic placement, except in very shallow locations.

4.4 Manufacture of micropipette electrodes

You will need:

- a microelectrode puller
- glass jars for electrode storage

The only materials you need, other than normal laboratory chemicals, are supplies of electrode glass. In order to fill with electrolyte properly, the glass must have some internal filament or septum to draw the electrolyte to the tip by capillary action. Borosilicate glass with an internal filament (from Clark Electromedical, Pangbourne, UK) should be used. The most popular size is 1.5 mm outside diameter, although 2 mm diameter glass makes slightly stronger pipettes; 1 mm glass is best left for intracellular electrodes (see Chapter 2).

Protocol 4. Manufacture of glass micropipettes

1. Fix a clean piece of electrode glass in the puller. Cleanliness of the glass, in terms of freedom from dust particles, is more important for pipettes than for solid conductor electrodes. If in doubt wash the glass in dilute nitric acid, followed by distilled water and then alcohol, and dry in a low temperature oven. Do not use detergent as this will leave residues which lower the surface tension of the electrolyte in the pipette and may impair proper filling.

2. Adjust the puller so that the pipette produced has a tip of about a micrometre when inspected under a microscope. Dimensions of less than a micrometre are too close to the wavelength of visible light to be clearly visible under most microscopes, so an alternative way of assessing the tip size is by the tip resistance. For extracellular work a resistance of 5–50 MΩ when filled with electrolyte is suitable. There are many different designs of puller and it is necessary to try a variety of settings of heat and pull strengths until the right combination is achieved. Very fine tips can be 'broken back' to a lower resistance.

Protocol 4. *Continued*

3. Fill the electrodes with 2 M sodium chloride solution by injecting the solution through a fine (100 μm or so) nylon tube attached to a hypodermic syringe. The nylon tube can be made from larger nylon tubing drawn out by hand under gentle heat (for example after warming in hot water). Choose a starting size that will fit over a large hypodermic needle; with practice a very fine tube can be obtained which can be pushed a long way into the pipette without damage.

4. If the electrodes are not to be used immediately it is a good idea to store them (unfilled) in a container filled with alcohol. The alcohol will gradually fill the electrodes but this can be replaced by electrolyte, when needed, by sucking out the alcohol from the shaft of the electrode, replacing with electrolyte and letting the tip fill by exchange diffusion. If kept in electrolyte for some time the tips can become eroded, or blocked by crystallization of the electrolyte. If kept in air dust particles are attracted by electrostatic forces to the tip, and become very difficult to remove.

5. Contact with the electrolyte is best made by a chlorided silver wire thrust into the stem of the pipette.

Key reference (7)

Although, in principle, suitable for DC-coupled amplifiers, micropipettes for extracellular recording are in practice used with AC-coupled amplifiers, unless there is some special requirement for DC-coupled recording. If DC recording is contemplated, special attention must be paid to the problems of tip potentials and the electrode potential of the chlorided wire; for spike recording AC amplification is perfectly adequate. Some researchers have found that 2 M NaCl, although it provides a lower resistance in the pipette, generates more noise than more dilute solutions. KCl is the electrolyte of choice for intracellular electrodes but can depolarize cells when used extracellularly. It is worth experimenting with 1 M or 0.5 M NaCl solutions to see which works best in any particular electrodes.

5. Other extracellular microelectrodes

There are a number of other types of extracellular microelectrodes that have been used at various times. The most popular are those where a platinum or platinum alloy wire, instead of a carbon fibre or tungsten wire, is used as the central conductor. The author has not tried making any of these other electrodes and so instructions as to their manufacture are not based on personal experience. For chronic recording the 'elgiloy' electrode seems to have many advantages and its fabrication has been described in detail (9–11). Tungsten electrodes have been used for recording from human peripheral nerve (12). However tungsten, unless platinized, is an unsuitable material for chronic contact with body fluids because it can form unstable oxide films on its surface which increase noise. Because of these problems, Jahnke developed a co-axial microelectrode made from a platinum wire embedded in a hypodermic needle, which is specifically designed for recording from human peripheral nerves. Details are given in his paper (13). For chronic recordings (over a period of days or more), tungsten electrodes have another problem; their very stiffness which makes them useful for acute stereotaxic placement makes them unable to move with pulsations of the brain and, therefore, liable to drift away from cells in the chronic preparation. To overcome this problem, flexible microwire electrodes can be used for single unit recording. These types of electrode were first described in 1958 (14), but recently have become much more popular. Good descriptions of the technique for making platinum wire and steel wire microelectrodes have been given in recent papers (15), (16). Steel microelectrodes made from etched insect pins and insulated with varnish have been used for a number of years and have been claimed to pick up more stable extracellular potentials than tungsten or glass microelectrodes (17). However, relatively few researchers use these electrodes and the unplatinized stainless-steel tip is likely to be electrically noisy.

6. Spike amplifiers

6.1 Preamplifiers

The electrode must be connected to a specialized amplifier (a 'preamplifier') in order to work properly. The preamplifier has to measure the voltage generated by the electrode without drawing more than a few picoamperes of current from the electrode. All amplifiers require some input current to activate them; but some, which use FETs (field-effect transistors) at the input need very much less than ordinary bipolar transistor inputs. Several makes of operational amplifier (op. amp.) are available with FET inputs; these are suitable for microelectrode preamplifiers. The key specifications for the op. amps are the bias current and voltage noise. The bias current is the current required to turn on the input FETs; the voltage noise should be of the order of a few nanovolts per square root frequency. At the time of writing the Burr–Brown Corporation (Tucson, USA) makes some of the best op. amps for use as front-end amplifiers with extracellular electrodes. They are in most good-quality electronic suppliers catalogues. An excellent all-round performer is the OPA 111 op. amp., and its dual version, the OPA 2111. These circuits have a bias current of 1–4 pA and a voltage noise of about 1 μV over the spike recording band. Very recently Burr–Brown have introduced a new pair of op. amps, the OPA 627 and OPA 637. These promise even lower noise at a slight cost in bias current (20 pA), and it should be interesting to see how they compare with the 111 types in microelectrode amplifiers.

There are two basic configurations for the headstage amplifier; a unity gain or non-unity gain, non-inverting format. The non-unity gain configuration

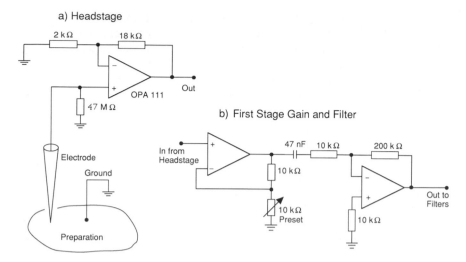

Figure 6. Headstage and first filter of extracellular amplifier. The low-noise FET-input operational amplifier (OPA 111) is used as a non-inverting amplifier with gain. The gain (× 10) is sufficient to raise the signal out of the amplifier noise region. The input shunt resistor (47 MΩ) should be used with tungsten or carbon fibre electrodes, but should be omitted for high impedance micropipettes. This resistor shunt removes DC potentials from metal or carbon electrodes and prevents input stage saturation during stimulus artefacts. The first stage gain and filter stages give further gain (say a further × 10) and also roll off low frequency signals outside the spike bandwidth.

gives better noise performance, but at the cost of sometimes saturating the output if large DC potentials are present on the electrode. In most cases a preamplifier gain of 10 is sufficient to bring the signal out of the noise level of the amplifier without risking saturation from large amplitude low frequency signals. A 47 MΩ or 100 MΩ resistor shunting the input to ground will remove DC and most low frequency signals from the input. (This resistor should be at least 100 MΩ when glass pipette electrodes are used.) *Figure 6a* shows the recommended circuit for a headstage amplifier suitable for tungsten or carbon-fibre electrodes.

6.2 Filters

The output signal from the preamplifier needs filtering to remove both high frequency and low frequency signals outside the spike bandwidth (approximately 300 Hz–8000 Hz). Because filters can be noisy circuit elements, it is a good idea to amplify the signal from the headstage before the main filters. A single-pole high pass filter will remove some of the low frequency interference and can provide gain as well. Use another OPA 111 in the configuration of *Figure 6b*. For simplicity of component values the main filters can be the Butterworth type. (These give a maximally flat signal amplitude in the pass-band, although they do produce some phase distortion.) *Figure 7* shows the circuit details for two

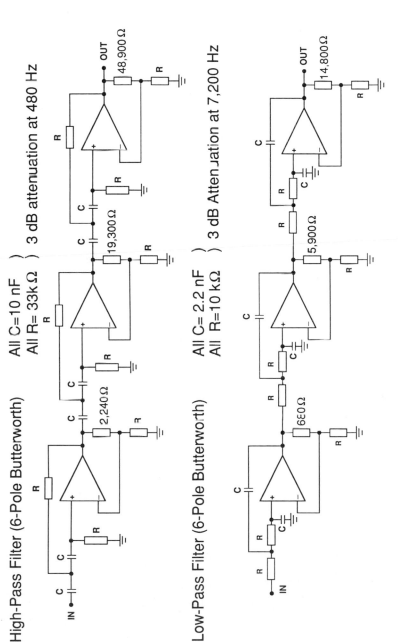

High-Pass Filter (6-Pole Butterworth)

All C=10 nF
All R= 33kΩ ⟩ 3 dB attenuation at 480 Hz

Low-Pass Filter (6-Pole Butterworth)

All C=2.2 nF
All R=10 kΩ ⟩ 3 dB Attenuation at 7,200 Hz

Figure 7. Active filters for extracellular amplifier. The values of these filters have been carefully chosen to optimize signal to noise ratio within the spike bandwidth. (Capacitors should be exact values. Resistor values can be produced by adding standard value resistors in series (e.g. 19 300 Ω = 18 000 + 1000 + 300 Ω)).

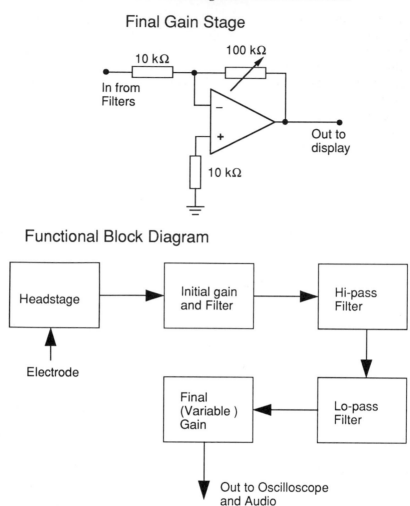

Figure 8. Final stage and block diagram of extracellular amplifier. The variable resistor here should be a front panel gain knob.

6-pole filters that give an attentuation of 36 decibel (dB) per octave outside the pass-band. The frequencies where the signal is attenuated by 3 dB (480 Hz and 7200 Hz) fit the requirements of most extracellular spike recordings and use easily realizable component values. For some purposes it may be necessary to retain more of the lower frequency components of the spikes; in this case the 10 nF capacitors in the high-pass filter can be changed to 22 nF or even 47 nF, but better signal-to-noise ratios will be obtained if the spikes are filtered to the pass-band specified in the diagram. Most commercial amplifiers do not have filters as good as these so there is something to be said for building your own, apart from cost!

Finally, the signal needs to be amplified to a point where a typical spike is large enough to be displayed on an oscilloscope or fed to an audio amplifier. This requires an overall gain in the whole system of about 10 000. The signal will already have been amplified by about 250 in the preamplifier and filter sections. A further variable gain stage of up to 50 can be incorporated in a final amplifier stage (*Figure 8*).

If line interference is still a problem, despite the use of the high-pass filter, a notch filter at line frequency may be needed. Consult a specialized text for this, for instance Horowitz and Hill (18).

6.3 Grounding and shielding

One of the main problems with extracellular recording is mains hum (pickup) and other electromagnetic interference. The high impedance of both the source microelectrode and of the preamplifier means that even small currents induced in the lead wire from the electrode can produce quite high voltages. For example if the electrode and amplifier impedances are taken in parallel and amount to 1 MΩ, then 1 nA of current induced in the wire will produce a 1 mV signal. In order to minimize this kind of noise, two rules should be adhered to:

Rule 1. Keep the lead wire from the electrode to the amplifier as short as possible. In some designs of commercial micropipette preamplifiers you can actually plug the electrode straight into the preamplifier. This is generally not possible with tungsten or carbon-fibre electrodes, but trouble can be minimized if the lead from the electrode is kept to 5 cm or less.

Rule 2. Place the preparation to be recorded from as close as possible to a large steel plate which should be connected to the ground or earth connection on the preamplifier.

The more massive the plate, the more helpful it will be at deflecting magnetic fields from the preparation, which normally cause most of the problem. Something 10–20 mm thick and at least 500 mm on each side (0.25 m²) is suitable. Faraday cages (large boxes of mesh around the preparation, connected to ground) do not necessarily remove interference from magnetic fields. The steel plate cannot remove interfering fields completely because the magnetic permeability of air is an appreciable fraction of that of steel, and so some proportion of any local fields will flow through the air above the plate.

Residual 50 or 60 Hz interference can be removed with special 'notch' filters which attenuate only the specified frequencies plus or minus a few per cent. Many commercial amplifiers offer notch filters as options. In fact, sine waves at line frequencies can be effectively removed by the high-pass filters described in Section 6.2. More pernicious are the problems arising from harmonics of line frequency. These arise from the rectifiers in the power supplies of nearby electronic instruments which convert the line voltage to 5, 12, or ± 15 V DC supplies. Power supply harmonics generate interference at 100, 200, and 400

(50 Hz line) or 120, 240, and 480 Hz (60 Hz line), and the higher of these frequencies will be within the pass-band of the spike amplifier. If problems with power supply harmonics occur, it is worth borrowing or hiring a spectrum analyser to pin down the precise frequency that is causing trouble, and then tracking down the offending instrument. When this has been done, you can either physically move the instrument further away from the preparation or try to screen the instrument in some way. A vertical grounded metal mesh between the instrument and the microelectrode may be effective. A last resort is to include a notch filter in the spike amplifier circuit tailored precisely to the interfering frequency.

Any line-powered equipment that has to be near the preparation, such as infusion pumps etc, should be grounded (earthed) to the ground terminal on the preamplifier to prevent ground loops. Ground loops occur when ground leads are connected in such a way that a loop or complete cyclic path for current exists. The loop may be several metres in circumference if it goes through ground leads to line-powered equipment. The loop thus formed can act as a tuned circuit to pick up line frequency magnetic field radiation and thus line hum. Use of a 'star' formation ground at the preamplifier input will minimize problems from ground loops. Everything near the preparation should be grounded to a single point at the preamplifier input **and to nothing else**. It is worth saying that electromagnetic interference and pick-up are the bane of extracellular recording, and most investigators will at some time in their lives spend frustrating hours trying to track down the source of some weird noise on their microelectrode. If despair is imminent, the book by Morrison (19) is of great therapeutic use. He manages to make some sense of the arcane principles of grounding and shielding, and the book can be understood without any mathematical background.

Spike signals, filtered and amplified, should always be monitored by both auditory and visual means. The ear is a surprisingly better discriminator of the shape of individual spikes than the eye. Feed the spike signal to a high-quality audio amplifier (the cost-effective way to do this is to use a commercial hi-fi amplifier). Spikes can be heard as cracks or clicks above the hiss of the baseline noise. This hiss can be very aurally fatiguing and some researchers will only listen to the clicks from discriminated spikes (see below). One way to remove the hiss, while retaining an ability to discriminate spikes aurally, is to clip the baseline noise with a circuit that cuts out all signals below a certain amplitude. This simple circuit (20) can be used to clip the input to the audio amplifier while visually monitoring the unclipped signal on an oscilloscope. A circuit diagram is given in *Figure 9*. The first potentiometer (lin pot 1) sets the gain, the second and third (lin pots 2 and 3) must be ganged together to create the clipped signal. The level they are set at determines the clip amplitude. The 200 k preset is used to balance the gain of the last amplifier. This circuit can be nicely realized with one TL 074 or equivalent quad FET op. amp.

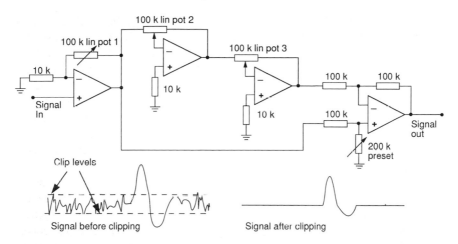

Figure 9. Noise clipping circuit. This circuit removes the low-amplitude components of the spike signal after filtering. It does not improve the signal-to-noise ratio but is useful for countering 'aural fatigue' when listening to spike signals for many hours during experiments. Lin pots 2 and 3 must be 'ganged' together so that as one attenuates the signal, the other augments it and vice versa.

7. Spike discriminators

As discussed in the Introduction to this chapter, the extracellular microelectrode is rather like a microphone in a crowd of people and, similarly, it can be difficult to discriminate which neuron is 'speaking'. The normal mechanism for solving this problem is to use a spike discriminator. This is a circuit that looks at the amplified and filtered signal from the electrode and generates a logic pulse whenever a transient in the signal satisfies a predetermined set of criteria, which usually involves amplitude and/or time course. An excellent review of discriminator circuits is given by Schmidt (21). The simpler kind of spike discriminator operates on amplitude alone. This is illustrated in *Figure 10*. The circuit has two voltage levels set by the operator. Spikes which have an amplitude intermediate between the two levels trigger a logic pulse. The difficulty with the circuit is in the upper or veto level. Some designs of circuit preset a fixed delay after the lower level is crossed. The second level must be crossed within this period to operate the veto. In other (better) designs the recrossing of the lower level triggers the logic pulse (as shown in *Figure 10*). Thus the spike cannot be discriminated until it is at least partly over. In order to see the leading edges of discriminated spikes on an oscilloscope a circuit to delay the spike by 0.5–1 ms is necessary.

More sophisticated spike discriminators use a time parameter, as well as voltage amplitude parameters, to distinguish between spikes of similar peak-to-peak amplitude but different time course. This is illustrated in *Figure 11*. *Figure 11a* shows two spikes which have different time courses but similar

23

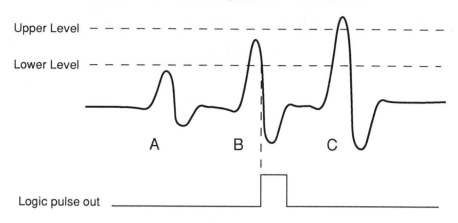

Figure 10. Spike amplitude discriminator. This shows the basic concept of an amplitude discriminator.

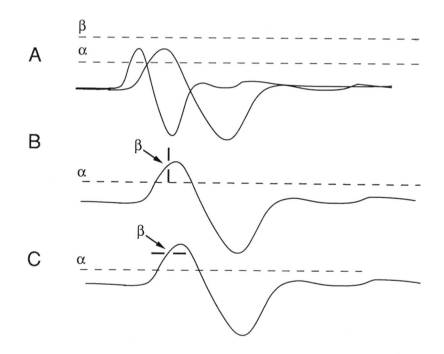

Figure 11. Spike time–window discriminator. A window system is a better way to discriminate spikes than a simple amplitude window, and either a voltage window (**11B**) or a time window (**11C**) can be used.

amplitudes, and are thus indistinguishable with a conventional discriminator. Two forms of time–window discriminator are shown in *Figures 11B* and *11C*. *Figure 11b* shows a fixed-time, variable amplitude window. To trigger the output pulse, the spike waveform must fall between the two voltage levels indicated at the time β, which is set by the operator at a fixed time after crossing level α. This form of window was developed by Bak (22). A commercial version is available (Bak Electronics Inc.). One advantage of this system is that two units can be cascaded to form a dual discriminator. The second form of time–window discriminator has a fixed voltage for the window but a variable open time. This is illustrated in *Figure 11C*. To be accepted the spike has to cross a fixed voltage at the level β between certain fixed times indicated by the gap in the β line. A system of this sort has been in use in the author's laboratory for several years (circuit details are available from the author). To be fully useful, the spike discriminator has to be used with a signal delay line so that the leading edges of discriminated spikes are visible. Signal delay lines (delays in the range 0.5–5 ms are needed) are available commercially; a simple circuit to construct a 12-bit delay line is given by Millar and Barnett (23).

Figure 12 shows the spike discriminator layout in use in the author's laboratory. The filtered spike signal is split, one channel goes straight to the spike discriminator and one goes through a precision delay line. The spike signal with multiplexed discrimination levels is displayed on one (non-storage) oscilloscope, triggered by a pulse generated when the spike crosses the first (α) level. This oscilloscope is used to decide which spikes are selected by adjustment of the

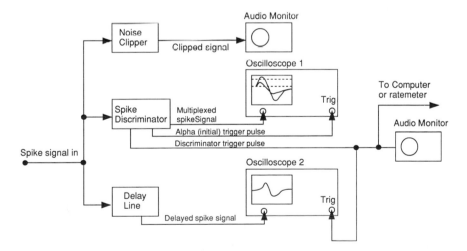

Figure 12. Spike discrimination system. This shows in block diagram form the apparatus necessary for accurate discrimination of single spikes from multi-unit data. At least two oscilloscopes are needed, one to display the multispike data and one to display the discriminated spikes. An analog delay line is also essential to see the whole of the discriminated spike.

discrimination parameters. The pulse produced at the end of the discriminated spike is used to trigger a second, digital storage oscilloscope, which is fed from the delay line. This second oscilloscope therefore displays *only* the discriminated spikes, including their leading edges. With most digital storage oscilloscopes you can 'save' a reference waveform and superimpose real-time waveforms on top. If one discriminated spike is saved, the goodness of the discrimination can be constantly monitored by the closeness of shape of the real-time spikes to the saved example. Finally, two audio monitors are used. One is fed from the spike signal via a baseline noise clipper to aid in the initial selection of spikes for discrimination; the second is fed from the discriminator and monitors the discriminated spikes only. Using two oscilloscopes and two audio monitors maximizes the investigator's ability to pick out the required spike signals from a multi-unit background.

Acknowledgements

I would like to thank Alan Ainsworth for discussions on methods for etching tungsten electrodes.

References

1. Hubel, D. H. (1957). *Science*, **125**, 549.
2. Merrill, E. G. and Ainsworth, A. (1972). *Med. Biol. Eng.*, **10**, 662.
3. Hellier, M., Boers, P., and Lambert, G. A. (1990). *J. Neurosci. Meth.*, **32**, 55.
4. Williams, G. V. and Millar, J. (1990). *Neuroscience*, **39**, 1.
5. Armstrong James, M. and Millar, J. (1979). *J Neurosci. Meth.*, **1**, 279.
6. Anderson, C. W. and Cushman, M. R. (1981). *J Neurosci. Meth.*, **4**, 435.
7. Millar, J. and Williams, G. V. (1988). *J. Neurosci. Meth.*, **25**, 59.
8. Geddes, L. A. (1972). *Electrodes and the measurement of bioelectric events*. Wiley-Interscience, New York.
9. Suzuki, H. and Azuma, M. (1976). *Electroenceph. Clin. Neurophysiol.*, **41**, 93.
10. Ashford, J. W., Coburn, K. L., and Fuster, J. M. (1985). *J. Neurosci. Meth.*, **14**, 247.
11. Hamada, T. (1985). *J. Neurosci. Meth.*, **14**, 253.
12. Vallbo, A. B. and Hagbarth, K.-E. (1967). *Electroenceph. Clin. Neurophysiol.*, **23**, 392.
13. Jahnke, M. T. (1982). *J. Neurosci. Meth.*, **6**, 335.
14. Strumwasser, F. (1958). *Science*, **127**, 469.
15. Mioche, L. and Singer, W. (1988). *J. Neurosci. Meth.*, **26**, 83.
16. Diana, M., Garcia-Munoz, M., and Freed, C. R. (1987). *J. Neurosci. Meth.*, **21**, 71.
17. Gottschald, K-M., Hicks, T. P., and Vahle-Hinz, C. (1988). *J. Neurosci. Meth.*, **23**, 233.

18. Horowitz, P. and Hill, W. (1980). *The art of electronics*. Cambridge University Press, London.
19. Morrison, R. (1985). *Grounding and shielding techniques in instrumentation*, (3rd edn). Wiley Interscience, New York.
20. Millar, J. and Barnett, T. G. (1983). *Electroenceph. Clin. Neurophysiol.*, **55**, 355.
21. Schmidt, E. M. (1984). *J. Neurosci. Meth.*, **12**, 1.
22. Bak, M. J. and Schmidt, E. M. (1977). *IEEE Trans. Bio-Med. Eng.*, **BME-24**, 486.
23. Millar, J. and Barnett, T. G. (1988). *Electronics and Wireless*, **94**, No. 1629, 648.

2

Intracellular recording methods for neurons

E. M. SILINSKY

1. Introduction

In the late 1920s, botanists interested in the behaviour of large plant cells pulled heated glass capillary tubing to give 1–2 μm tips, filled the resultant pipettes with electrolyte solutions, and impaled *Valonia, Chara*, and *Halioclystis* cells to study their transmembrane electrical properties (1). The use of such pipettes in animal cells soon followed when Cole and colleagues (2) recorded intracellularly from embryonic heart cells in tissue culture. The creation of the true microelectrode (generally defined as possessing a tip diameter < 1 μm) is frequently attributed to Gerard, Graham, Ling, and colleagues in the 1940s (3), who first described the conventional double pull technique for creating electrodes of 0.5 μm or less. The methods of Gerard and his associates also formed the link between the earlier artistry of those who pulled microelectrodes by hand and our present technological era of microprocessor-controlled electrode pullers. The diligence of these electrophysiological pioneers, coupled with post-war improvements in electronic circuitry, laid the groundwork for the next 30 years in which intracellular microelectrodes dominated the field of electrophysiology.

The recent development of patch clamp methods, a logical extension of intracellular recording methods, has superseded some of the need for intracellular microelectrode recording (see Chapter 3). There are numerous instances, however, in which intracellular recording techniques are the most appropriate ways for investigating the physiology of neurons. This is especially true for studies of adult vertebrate neurons and synapses *in vitro*. Indeed, in most instances, intracellular microelectrode recordings provide reliable estimates of physiologically-functional transmitter secretion with minimal damage to the cell and without producing drastic changes in the levels of intracellular messenger substances.

This chapter considers the basic principles of intracellular recording from neurons. The chapter begins with a description of an actual experiment on intracellular recording of synaptic potentials, thus providing an overview of how to study neuronal communication initiated by the presynaptic nerve. The experiment is then considered from the practical and technical perspectives of microelectrode and amplifier considerations. Finally, a protocol for an experiment in which the membrane potentials are controlled directly by the experimentalist under current clamp and voltage clamp are provided. Due to the commercial availability of superior amplifiers designed exclusively for electrophysiologists (for example the Axoclamp series from Axon Instruments), only a limited discussion of theoretical instrument design is provided. Further details of the electrophysiological methods may be found in some excellent books (4–7) and in original papers (8–10).

2. Synaptic control of membrane electrical behaviour: description of a typical experiment

The strongest justification for using single microelectrodes for intracellular recording is in the study of synaptic communication between a chemically-transmitted neuron and its receiving cell. Indeed, the most reliable information with minimal changes in neuronal physiology may be obtained from such studies, without the need for voltage clamping (10). The experiment on synaptic potentials begins with a preparation (*Figure 1*), in this instance lumbar sympathetic chain ganglia from frog, pinned in a tissue bath and superfused with oxygenated salt solution. The presynaptic nerves are coaxed into a suction electrode for stimulation (⎍). A microelectrode is then fabricated, filled (with 3 M KCl), placed inside its holder (usually of Perspex and containing a Ag/AgCl wire or pellet), and the holder in turn inserted into the headstage of the preamplifier. The output of the headstage is fed into the main power supply of the preamplifier, which is equipped with various circuits for providing digital monitoring of membrane potentials (-69.8 mV in this experiment), DC offset, capacitance compensation, and output connectors. The outputs are delivered to a cathode-ray oscilloscope and other recording and display devices, such as a computer, magnetic tape recorder, and pen recorders. Procedurally, the electrode/headstage combination is moved under visual control using a micromanipulator into the bath. At this juncture the preamplifier measures a difference of potential between the microelectrode and the Ag/AgCl reference electrode connected to earth. This potential is cancelled by DC offset. The cell is then impaled with the microelectrode, recording the resting membrane potential (*Figure 1*, V_m) between the inside of the cell and the reference electrode outside the cell. Depending upon the preference of the particular investigator, the progress of the impalement may be monitored from the digital readout on the preamplifier, the oscilloscope screen, the pen recorder, or the computer monitor. It is from this position of V_m (generally -50 to -70 mV in vertebrate neurons) that the experiment begins and the investigator searches for interesting deflections. In this experiment on synaptic potentials, the nerve is stimulated and an excitatory synaptic potential (esp) occurs due to the release of acetylcholine (ACh) from the presynaptic nerve on to the neuron which is being recorded (trace **a**). The depolarization associated with the esp is due to the inward movement of Na$^+$ through non-selective cationic channels *gated by*

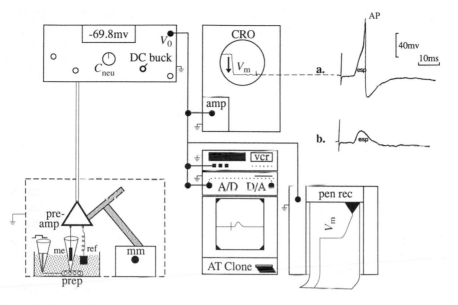

Figure 1. Electrophysiological apparatus for intracellular microelectrode recordings. Abbreviations are as follows: prep, neuronal preparation (in this experiment, frog sympathetic chain ganglia); me, microelectrode; ref, reference electrode; preamp, preamplifier headstage; mm, micromanipulator. On the main part of the amplifier, note digital meter for membrane potential measurements (V_m, −69.8 mV), C_{neu}, negative capacitance; DC buck, provides DC offset of extracellular potentials prior to impalement. V_o provides outputs to (i) an amplifier (amp) of a cathode ray oscilloscope (CRO), (ii) a video cassette recorder (vcr), (iii) an analog to digital (A/D), digital to analog (D/A) converter (which in turn delivers its signal into a microcomputer (AT Clone), and a pen recorder (pen rec). Experimental traces shown on the right are from an unpublished experiment. In (a), note the excitatory synaptic potential (esp) produced in response to presynaptic nerve stimulation (⎍) is suprathreshold and generates an action potential (AP). In (b), drug X reduces the esp below threshold for AP generation. Drug X is ATP (500 µM), which is a presynaptic inhibitor of ACh release in this preparation (23). Spikes prior to esps in (a) and (b) are stimulus artefacts. See Appendix 1 of this chapter for additional discussion of the experiments.

the transmitter *ACh* in the impaled cell. Also observed is a superimposed action potential produced by Na⁺ and K⁺ movements through separate *voltage-gated channels*. In the experiment (shown in trace **b**), a substance X produces a decrease in the esp so that it no longer reaches threshold for action potential generation. Changes in esps are quantified by digitizing the data using an analog to digital converter (A/D) and then feeding the digitized data into a microcomputer. Based upon this experiment, substance X could act *presynaptically*, to inhibit the release of ACh or *post-synaptically* to reduce the effects of ACh. Ways of distinguishing between these alternatives are provided in Appendix 1 of this chapter.

This description will hopefully provide the reader with a brief overview of the methods used to study neuronal synaptic communication. Some of the more technical and practical aspects of intracellular microelectrode recording will now be considered.

3. Practical and theoretical aspects of the typical experiment

3.1 Practical aspects

3.1.1 Microelectrode fabrication, filling, and connections

Modern microelectrode pullers, whilst eliminating many of the artistic aspects of pulling microelectrodes, have allowed electrophysiologists to concentrate their efforts on studying the behaviour of the excitable cells. Indeed, the *fabrication* of a microelectrode is currently a perfunctory task (see Chapter 3, Section 5.2). The general rule of thumb for intracellular recording is that the tip diameter should be 1/10 to 1/100 of the average cell diameter to prevent the microelectrode from damaging the cell (11).

Before 1975, a number of rather tedious methods were used to fill the tips of newly-fabricated electrodes. Filling is now almost exclusively performed by the glass-fibre method originated by Tasaki and colleagues (12). Specifically, commercially available microelectrode glass may be purchased with fine glass fibres affixed inside so that the electrode may be filled by injecting the solution with a syringe, the fibre acting as a hydraulic conduit allowing capillary action to fill the tip. This technique also allows the investigator to be unconcerned about air bubbles in the shank. Conventionally, the most frequently used filling solution is KCl (2–3 M); electrodes filled with KCl have low resistances and minimal liquid junction potentials (see below). KCl-filled electrodes are generally not used to study chloride currents as the high concentrations of KCl can change the intracellular Cl^- concentration. Other filling solutions include potassium citrate or acetate (to improve current passing dynamics of the electrodes at the expense of higher resistance), caesium (Cs^+) salts (to inhibit K^+ currents), or mixtures of different solutions, often containing possible substances to be injected into the cell, such as GTP analogues or Ca^{2+} chelators.

3.1.2 Methods of cell impalement

The history of intracellular recording is replete with anecdotes about impalement methods. One of the earliest impalements of sympathetic ganglia was produced by Dr Gary Blackman stomping on the floor on the Edinburgh laboratory when the electrode was resting on the ganglion cell (B. L. Ginsborg, personal communication and ref. 13). Prior to the advent of newer preamplifiers, the general method of impaling cells was to press the electrode against the cell and tap gently on the antivibration mounting or another part of the experimental rig (for example the fine adjustment of the micromanipulator, ref. 14). The most widely employed method nowadays is first to press the electrode on the cell and then 'buzz' the tip into the cell using a high amperage, oscillatory current pulse for a very brief period. Some amplifiers (for example Axoclamp) come equipped with buzz switches for just this purpose. This effect is thought to be due to a breakdown of the membrane dielectric locally in the region of high current density near the electrode, thus allowing the electrode to enter the cell. Once placed intracellularly, the stability of impalements are greatly improved in higher Ca^{2+} concentrations. Thus, if Ca^{2+}-free solutions are to be employed with intracellular microelectrodes, it is suggested that the experimenter first obtain a stable recording in normal or slightly elevated (for example 5 mM) Ca^{2+} solutions before reducing the concentration of Ca^{2+}. Preamplifiers, to the headstage of which the microelectrode and holder is usually attached, are generally at unity gain ($\times 1$ amplification) but have high input impedance as a consequence of field effect transistor (FET) circuitry. Such preamplifiers will be discussed in the theoretical section below.

3.1.3 Display, recording, and analysis

In the past, when electrophysiological data was photographed directly from the oscilloscope screen, the possession of a high quality oscilloscope was essential for hard copy of the experimental results. Pen recorders and, more recently, computers have superseded the oscilloscope in this regard, and almost any oscilloscope will suffice as long as plug-in amplifiers with suitable gain (for example 1 mV/division) are present. For example, many studies of slowly changing ionic currents employ pen recorders (for example Gould Instruments) for data collection. For computer-assisted data acquisition and analysis, the vast majority of experimenters use IBM-PC based analog/digital (A/D) and digital/analog (D/A) interfaces with associated software in conjunction with an XY plotter for producing hard copy (see Appendix 2 of this chapter for suggestions concerning specific items of equipment).

3.2 Theoretical aspects

3.2.1 Preamplifiers

As oscilloscopes frequently contain very high gain amplifiers, why can the signal from the microelectrode not be fed directly into the oscilloscope without the need for a high-input resistance preamplifier (preamplifiers rarely add any amplification at all as they have unity gain)? The purpose of this section is to address this issue.

A simplified equivalent circuit of the microelectrode and its amplifier is shown in *Figure 2*, with R_{me} representing the resistance of the microelectrode, V_m representing the cell membrane potential and R? the input resistance of the pre-amplifier. If $R_{me} = 20 M\Omega$ (typical for an intracellular recording microelectrode in many cells), and the input resistance of the preamplifier is 1 MΩ (which is typical for an oscilloscope amplifier) then, as the circuit is a conventional voltage divider, only $1/20 + 1$, that is 1/21 or about 5% of the measured membrane potential will be recorded at the input terminals of the amplifier. In contrast to an oscilloscope amplifier, the input impedance of even a 17-year-old WP Instruments high input impedance preamplifier is 20 000 MΩ. Under these conditions:

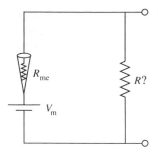

Figure 2. Simplified equivalent circuit of microelectrode resistance (R_{me}) and input resistance of preamplifier (R?). V_m, membrane potential.

$$\frac{2\times10^{10}}{2\times10^{10}+2\times10^7}=99.9\%.$$

Thus, all but one tenth of one per cent of the true signal voltage will be recorded with such a high input impedance preamplifier. Some headstages for the Axoclamp 2A, available for use with high resistance ion selective microelectrodes, possess input resistances as high as $10^{14}\,\Omega$.

3.2.2 Electrode potentials, resistances, and capacitances

The first and most important interface between the cell and the experimenter is the microelectrode. It is thus necessary to discuss in some detail the passive circuit element properties of the microelectrode. The potentials, resistances, and capacitances of the microelectrode and its associated preamplifier circuitry are illustrated in *Figure 3*.

Beginning with *potential* measurements, the membrane potential of the neuron is only one of many contributing factors to the difference of potential measured between the intracellular recording electrode and the indifferent Ag/AgCl reference electrode (ref). Potential differences which can reach hundreds of millivolts exist even with the microelectrode present in the extracellular fluid. These potentials are due to differences in electrical behaviour that occur at interfaces between solutions of different composition or at metal–electrolyte interfaces. For example, the fluid in the electrode is generally different from the extracellular fluid and, if the mobilities of ions differ, then a diffusion or liquid junction potential (E_{ljp}) develops. Two other potentials are at metal (Ag/AgCl)–electrolyte junctions, the first where the metal wire inside the electrode contacts the 3 M KCl filling solution, and the second between the reference electrode and the extracellular fluid. These potentials are designated $E_{Cl'}$ and $E_{Cl''}$, respectively, in *Figure 3a* to emphasize the fact that Ag/AgCl electrode potentials reflect the activity of chloride ions. These Cl^- activity potentials may be as large as several hundred millivolts.

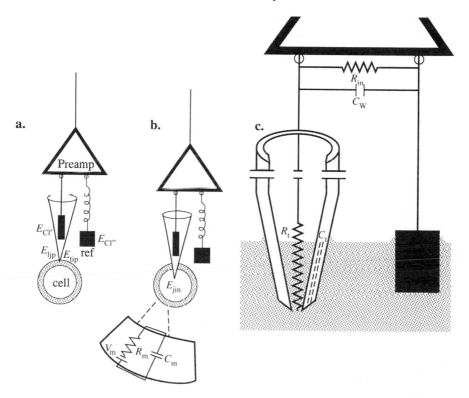

Figure 3. Potentials, resistances, and capacitances associated with the microelectrode and the preamplifier. (a) and (c), electrode in extracellular fluid outside of cell. (b) Electrode is intracellular. Symbols are as follows: in (a) E_{ljp}, liquid junctional potential between the electrode solution (3 M KCl) and the extracellular solution; E_{tip}, tip potential of the microelectrode; $E_{Cl'}$, potential between the Ag/AgCl wire in the microelectrode and 3 M KCl (reflects chloride ion activity); $E_{Cl''}$, potential between Ag/AgCl reference electrode and extracellular fluid (also reflects chloride ion activity). In (b) E_{jin}, liquid junction potential between electrode solution and cytoplasm; V_m, membrane potential; C_m, membrane capacitance; R_m, membrane resistance. In (c): R_t, microelectrode tip resistance; C_t distributed capacitance in the wall of the microelectrode separating 3 M KCl from the bathing extracellular fluid; R_{in} input resistance of the preamplifier. C_w, capacitance of the wiring and of the input of the preamplifier.

Another extracellular potential, the *tip potential* of the electrode (E_{tip}), merits special consideration. E_{tip} is defined as the potential that disappears when a high resistance microelectrode has its tip fractured; it occurs because capillary glass pulled to very fine tips develops unusual ion exchange properties. Specifically, the electrode tip acquires a layer of fixed negative charges while the adjacent solution compensates by acquiring a thick layer of cations, so much so that a cationic plug may develop and hinder anion movement across the tip. Tip potentials are increased as the electrode resistance rises (i.e. the tip becomes smaller). E_{tip} is generally less than 10 mV for electrodes of 0.5 μm diameter.

35

To summarize these extracellular potentials, the measured potential difference between the extracellular microelectrode and reference electrode can be derived thus:

$$E_{measured} = E_{tip} + E_{ljp} + E_{Cl'} - E_{Cl''}. \tag{1}$$

These extracellular potentials do not constitute a major impediment to intracellular studies as they are cancelled by the variable battery supplied with all commercially-available preamplifiers (*Figure 1*, DC buck). Thus after placing the microelectrode in the extracellular fluid, these potentials are eliminated by the offset control so that the voltage monitor (oscilloscope or digital voltmeter) reads 0. The cell is then impaled and the general assumption made that the newly measured potential difference represents only the membrane resting potential. If neither E_{ljp} nor E_{tip} change when the cell is impaled, then this assumption would be accurate. Unfortunately, the liquid junction potential changes when the electrode enters the cell because 3 M KCl in the microelectrode now is presented with a solution of different composition from the extracellular bathing solution, namely the cytoplasm. This difference in junctional potential, $E_{\Delta ljp}$ may be as large as 15 mV and is independent of electrode resistance (15,16). Tip potentials may also change as the electrode penetrates the cell. The cause of the change in tip potential ($E_{\Delta tip}$) is unknown, although it is perhaps due to interactions with or even clogging by intracellular proteins. Thus after eliminating the extracellular potentials and impaling the cell the measured intracellular potential difference is given by the expression:

$$E_{measured} = V_m + E_{\Delta ljp} + E_{\Delta tip}. \tag{2}$$

Given the above equivalence, how accurately may we measure membrane potentials? In practice, the use of 3 M KCl makes $E_{\Delta ljp}$ small as K^+ and Cl^- have approximately equal mobilities, thus the charges do not separate greatly and liquid diffusion potential differences may be minimized to less than 10 mV (see ref. 5, 168–172; and ref. 4, Chapter 3 for more details). Generally, liquid junction potential differences tend to make the resting membrane potential appear less polarized than the true level. Liquid junction potentials are a very important consideration with patch electrodes (see Chapter 3, Section 5.6). Problems concerning tip potentials are currently of diminished importance as the fibre method for filling microelectrodes greatly reduces the tip potential (<3 mV for 15–30 MΩ electrodes (12)). Tip potentials may also be minimized by choosing electrodes of low resistance and those made from quartz glass.

Thus, under these ideal conditions, $E_{measured}$ equals V_m. However, with high resistance electrodes (for example 100–200 MΩ often used for some cells such as enteric neurons), resting potentials may be difficult to quantify. For example, small leakage currents around the electrode membrane seal when flowing across a high resistance electrode may lead to spurious measurements of V_m. The reader is thus advised to exert some caution in assuming that the measured potential upon impaling a cell is an accurate reflection of the transmembrane resting potential.

It is now necessary to address the issues of resistance and capacitance. Apart from those passive circuit elements of biological interest, namely the membrane resistance (R_m) and capacitance (C_m) (*Figure 3b*), the electrophysiological recording system possesses resistance and capacitance of its own (*Figure 3c*). The main *resistances* of concern are those due to the small tip of the microelectrode, R_{tip} and the input resistance of the preamplifier, R_{in} (see also *Figure 2* and related discussion). (Wire, fluid, and interface resistances are relatively small and may be ignored.) With respect to *capacitance*, whenever two conducting surfaces are separated by an insulator, capacitance is present. Capacitance makes all voltage changes sluggish as time is required to charge the capacitor. Specifically, the time for the voltage to rise to 63% of its maximum value (the time constant, τ is equal to the product of the resistance and capacitance):

$$\tau = R \times C. \tag{3}$$

In the circuit shown in *Figure 3c*, significant tip capacitance (C_t) is distributed along the insulating electrode wall separating the pipette filling solution and the bathing solution. Some minor capacitance is also introduced by wiring to the input terminals of the amplifier (C_w). The net effect of these capacitances, especially the large distributed capacitance along the electrode wall, may prevent smooth decay of potentials recorded by the microelectrode. The fast and slow stages of decay of the voltages across a microelectrode are best observed by monitoring the output of the headstage of a single electrode voltage clamp amplifier (see section below).

The resistive and capacitance elements add noise to the recording system and prevent the electrode from faithfully following potential changes. How may one reduce their magnitude? To reduce resistance in the circuit, use the lowest resistance electrodes compatible with stable intracellular recordings from undamaged cells. The RMS voltage noise across a microelectrode is proportional to the square root of the resistance in Ω (the proportionality factors being absolute temperature, Boltzmann's constant, and the bandwidth of the measuring system). This is called Johnson noise and for a 10 MΩ electrode and an amplifier bandwidth extending from 0 to 10 kHz, this amounts to 150 µV peak-to-peak. For a > 100 MΩ electrode used to study many smaller neurons, the peak-to-peak noise would be approaching the mV level. Some investigators bevel their microelectrodes to produce lower resistances with sharper tips, thus reducing electrode noise yet allowing for successful impalements (see ref. 4, p. 24). To reduce circuit capacitance, first make the wiring as short as possible—this is why headstage amplifiers are mounted on the micromanipulator and the microelectrode holder plugs directly into the headstage. Secondly, keep the tissue bath volume low as less distributed tip capacitance occurs. Thirdly, coat the electrodes with an insulating material, for example Sylgard or another silicone polymer, which prevents creep of the bathing solution up the tip of the electrode (see Chapter 3, Section 5.3). Such coating also increases the size of the insulation between conducting surfaces thus reducing the capacitance. (C is proportional to 1/distance between insulating surfaces. Use of insulating coating reduces C by increasing the size of the insulator, thus increasing the distance between the conducting surfaces.) Finally, make use of a knob on every preamplifier called negative capacitance. Negative capacitance supplies the rapid missing parts of the sluggish voltage trace caused by the RC circuit, that is by feeding back a differentiated signal to neutralize the capacitative loss (see knob C_{neu} in *Figure 1*). Appropriate capacitance neutralization is illustrated in *Figure 4*: **a** shows the input voltage, **b** shows the sluggish microelectrode response with no compensation, **c** shows an adequately compensated tracing, and **d** shows excessive negative capacitance. For further details of other methods to reduce stray C and R (for example bootstrapping, driven shields, etc. see refs 4 and 5).

Appropriate earthing should also be employed to reduce noise in the electrophysiological apparatus (4,10). Briefly, all ground wires coming from the equipment chassis (and from a Faraday cage around the preparation and electrode, if needed—see dashed line in *Figure 1*) should be joined at one *and only one* point (chassis ground) (see Chapter 1, Section 6.3). For further details of other aspects of noise reduction (for example moving motors away from the rig, using DC power supplies for microscopes: see refs 4,10). Finally, to ensure mechanical stability, an antivibration mounting is necessary on which to affix the preparation and micromanipulator (see Appendix 2 this chapter and ref. 4 for details). From this limited foray into the practical and technical aspects of microelectrodes and their connections, there will now be a discussion of measurements of cell electrical behaviour.

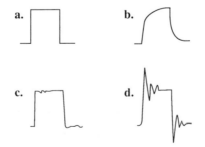

Figure 4. Use of negative capacitance to neutralize input capacitance and improve the voltage response. (a) input voltage, (b) response of electrode without compensation, (c) proper capacitance neutralization, and (d) excessive use of negative capacitance.

4. Extrinsic control of membrane electrical behaviour

4.1 Voltage recording under current clamp

In the experiment shown in *Figure 1*, the investigator was at the mercy of the presynaptic neuron. The remainder of this chapter focuses on conditions in which the experimenter takes a greater role in altering the electrical behaviour of the neuronal membrane in order to probe its intricacies.

A common way of studying the membrane properties of chemically and voltage gated ion channels is to use a single microelectrode in conjunction with a bridge circuit for both passing current and measuring changes in membrane potential. In the past, when passive elements were employed, the circuit was known as the Wheatstone bridge. Currently, active bridges, that is operational amplifiers, are used in bridge studies with single electrodes under current clamp.

Procedurally, constant current pulses (*Figure 5a*) are passed through the microelectrode *before* impalement, and capacitance neutralization is used to sharpen the voltage response prior to attempting to balance the bridge (see above). *Figure 5b* shows the microelectrode voltage (*V*) after capacitance neutralization, but before bridge balance. The bridge is then balanced by rotating a calibrated potentiometer until the voltage drop across the electrode arm is

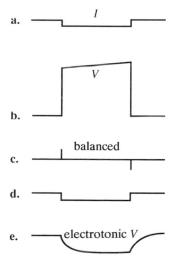

Figure 5. Use of a bridge circuit in current clamp. Trace (a) shows constant current pulse; (b) shows uncompensated voltage response of the microelectrode to the current pulse; (c) shows appropriate compensation to produce bridge balance; (d) shows overcompensation. Traces b–d were extracellular responses. The cell is impaled after bridge balance and the electrotonic potential across R_m and C_m recorded (e). The steady phase of the electrotonic potential provides a measure of the input resistance of the membrane. Specifically, the current I and electrotonic potential are measured and R_m determined from Ohm's law as R_m, electrotronic V/I. C_m may then be estimated from the time constant of the membrane (τ_m, the time for the voltage to reach 63% of the steady level) as $\tau_m = R_m C_m$, *see* equation 5 in text.

eliminated (*Figure 5c*). Under conditions of bridge balance, the potentiometer provides the experimenter with an estimate of the resistance of the electrode. If the bridge is overbalanced, a negative potential change occurs due to the active bridge circuit passing current (*Figure 5d*); this condition is to be avoided. Once the bridge is balanced (*Figure 5c*), the cell is impaled and an additional voltage change occurs across the input resistance R_m and capacitance C_m of the membrane. This potential is termed the electrotonic potential (*Figure 5e*) and represents the biological behaviour of interest. If a particular perturbation decreases R_m, that is increases membrane conductance by opening additional ion channels, then the steady phase of the electrotonic potential is smaller (the steady phase represents current across the membrane resistance as the capacitance is already fully charged—see below). Specifically, by Ohm's Law, $E_{electrotonic}$ is equal to $I \times R_m$ or $E_{electrotonic}$ is equal to I/G_m where G_m is the input conductance of the membrane (the reciprocal of R_m). Thus, in the face of constant current a decrease in $E_{electrotonic}$ must occur due to a decrease in R_m (increase in G_m).

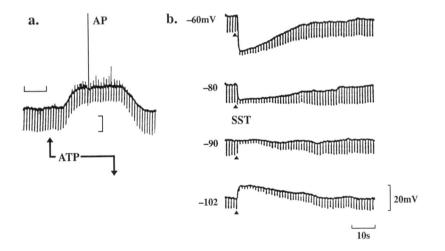

Figure 6. Examples of experimental results using the bridge circuit in mammalian neurons. (a) Shows results from cultured guinea-pig coeliac neurons, in which ATP (200 nM, applied between arrows) produced depolarization associated with a conductance increase. Note a decrease in the size of electrotonic potentials (downward-going spikes) indicating a conductance increase (resistance decrease) to Na^+ and other cations (see text). Upward spikes arising from depolarized trace are action potentials (AP). Calibration (15 mV, 10 s). (b) Shows results from guinea-pig submucous neurons *in vitro* in which somatostatin (SST) produces hyperpolarization from the resting level (-60 mV); the SST effect is also associated with a conductance increase, but as the reversal potential is near -90 mV (the potassium equilibrium potential), the conductance increase is to potassium ions and not to sodium. Membrane potentials prior to the application of SST (applied by puffing at the arrow) shown on left (reprinted with permission from 17).

An experiment in which ATP (200 nM) depolarizes guinea-pig coeliac neurons and increases its conductance is shown in *Figure 6a*. Note the decrease in the size of the resistive component of the electrotronic potential which, at this recorder speed, appears as a spike. Another experiment from the work of Mihara, North, and Surprenant (17) on neurons in the guinea-pig small intestine is shown in *Figure 6b*. Note that in *Figure 6b*, the somatostatin (SST—arrow) also increases the conductance of the membrane, but, in contrast to the effects of ATP shown in *Figure 6a*, hyperpolarizes the neuron from its resting level (-60 mV, upper trace). The membrane conductance increase is to potassium ions, as the potential change reverses near the potassium equilibrium potential (approximately -90 mV). For quantitative purposes, if the resting membrane has a linear current voltage relationship, then it is possible, by tracing the envelopes of the electrotonic potentials for a single response, to measure the reversal potential for the applied transmitter (for mathematical details see Ginsborg, House, and Silinsky, ref. 18).

Experiments such as those illustrated in *Figure 6* aid in identifying: (i) if a putative neurotransmitter could actually be involved in the synaptic response (by comparing the reversal potential for exogenously applied and real transmitter) and (ii) the nature of the ions that move through the channel.

Other variants of the current clamp technique include the discontinuous current clamp (see below) and two electrode current clamp. If the cell may be impaled with two electrodes, however, it is best to use the two electrode voltage clamp to study the membrane properties (see below).

4.2 Current recording under voltage clamp

4.2.1 Overview

The voltage clamp amplifier is, in essence, a *high gain negative feedback device*. Simply, one input of the clamp amplifier (*Figure 7*) records the signal from the unity gain preamplifier ($\times 1$) and at the other input the voltage is commanded (V_c) to remain at some level, for example the resting potential. If excitatory substances such as ATP or ACh are then applied, the membrane potential tries to change but cannot because the clamp amplifier passes current to prevent the potential from changing (I_m). This current represents the ATP or ACh induced membrane current and its amplitude is unencumbered by depolarization of the membrane (see below). This figure illustrates the two electrode voltage clamp (TEVC) in which one microelectrode (V) samples membrane potential and a second microelectrode (pass I_m) passes current.

Why is the voltage clamp so valuable? In the experiment discussed above (*Figure 6a*), ATP receptor activation depolarizes neurons and increases membrane conductance. As depolarization in itself increases membrane conductance (by opening voltage-sensitive Na^+, K^+ and Ca^{2+} conductances in the membrane), it could be argued that membrane depolarization, in itself, opens voltage-sensitive conductance pathways, and ATP could be acting via a mechanism

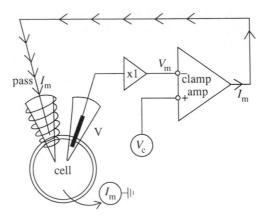

Figure 7. Basic principles of voltage clamping as illustrated by the two electrode voltage clamp (TEVC). V, voltage recording electrode; I_m, membrane current; pass I_m, current passing electrode; $\times 1$, unity gain high input impedance preamplifier; V_m, membrane potential; V_c, command potential; clamp amp, voltage clamp amplifier.

unrelated to the opening of ion channels, for example by changing the activity of pumps or transporters which in turn depolarizes the membrane and opens voltage-gated ionic channels. Such an argument could be excluded by voltage clamping. In addition, quantitation of conductance increases to depolarizing transmitters is better performed under voltage clamp for two main reasons. Firstly, regenerative effects of depolarization on ionic channel gating occur. For example, ATP-induced depolarization opens voltage-gated Na^+ channels, which further depolarizes the membrane and thus further opens Na^+ channels until regenerative all-or-none action potentials (AP) occur. Secondly, the size of the depolarization is not linearly related to the conductance change (Δg_{ic}) produced by a particular concentration of transmitter.

Specifically the relationship is defined as:

$$\textit{Depolarization by transmitter} = \left(\frac{\Delta g_{ic}}{\Delta g_{ic} + G_m}\right)(V_m - E_{ATP}). \qquad (4)$$

where Δg_{ic} is the conductance of the ion channel added by the transmitter and G_m is the membrane input conductance.

The linearity of the anticipated relationship between transmitter-induced conductance change (Δg_{ic}) and depolarization is skewed by two factors in Equation 4. Firstly, the fact that the expression is really a rectangular hyperbola (except when Δg_{ic} is very much less than G_m), and secondly, changes in driving force. Driving force is defined as the difference between the resting potential (V_m) and E_{ATP}, the equilibrium potential for the ions that flow through the ATP-gated channels; that is driving force is the ($V_m - E_{ATP}$) term in Equation 4. Thus the depolarization produced by an individual increment in conductance

Δg_{ic} is dependent upon the driving force. If V_m is close to V_{eq}, then large increases in Δg_{ic} could produce only small changes in membrane potential when compared with a similar conductance change at more polarized V_m and thus at a larger driving force.

If the membrane potential could be 'clamped' at the resting level, ACh or ATP then applied and an inward (depolarizing) current associated with a conductance increase measured, this would provide strong evidence that ACh or ATP activates a membrane ion channel and directly increases membrane conductance without the need for depolarization. Under voltage clamp, I_m is equal to $\Delta g_{ic} \times$ driving force. Thus no change in driving force occurs and changes in I_m are directly related to the conductances of the transmitter activated ionic channels in voltage clamped cells. Concentration–response curves may thus be more reliably constructed under voltage than current clamp, as voltage clamp allows measurements of transmitter-induced currents to be made apart from secondary considerations of changes in driving force or non-linear relationships between Δg_{ic} and depolarization.

4.2.2 Theoretical aspects

One of the most important properties of a good voltage clamp is the ability to achieve a *high clamp gain* (10^8 or greater). To understand this, it is best to examine the equivalent electrical circuit of *Figure 8* under conditions in which the voltage output (V_o) to the current passing microelectrode is a battery with V_o supplied by a voltage clamp amplifier. In either case, by Kirchoff's Loop Law (i.e. the sum of voltages around a closed loop is equal to zero) applied to the circuit in *Figure 6* V_o is equal to the sum of IR_e and V_m. For the case of the battery, if V_o is equal to $E_{battery}$, V_o would never reach V_m as it is decreased by the IR_e voltage drop across the microelectrode. If V_o is supplied by a clamp amplifier, then $V_o = GAIN\ (V_m - V_c)$ (see refs 6 and 7). Thus $GAIN(V_m - V_c) = IR_e + V_m$. Rearranging and solving for V_m

$$V_m = V_c \frac{GAIN}{GAIN+1} - IR_e \frac{1}{GAIN+1} \qquad (5)$$

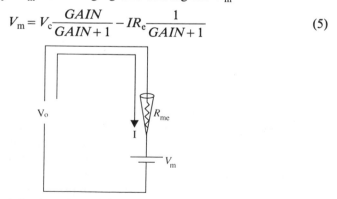

Figure 8. Simplified circuit for the voltage delivered to a current passing electrode (V_o) in a voltage clamp experiment. See text for consideration of differences when V_o is provided by a simple battery and when V_o is provided by a voltage clamp amplifier of high gain, R_{me}, resistance of the microelectrode; V_m, resting membrane potential.

As the gain approaches very large values, $GAIN+1$ is equal to $GAIN$ and the first term after the $=$ sign ($V_c \dfrac{GAIN}{GAIN+1}$) approaches V_c. Large $GAIN$ values will also cause $1/GAIN+1$ in Equation 5 to approach zero, thus the second term after the $=$ sign to approach zero. High GAIN therefore eliminates the voltage drop across the microelectrode, (IR_e) and causes V_m to be equal to V_c. This is the ultimate aim of the voltage clamp! The other property of importance for voltage clamping is *good spatial control*, that is, space clamp. To understand space clamp, it is important to consider the mathematical description of the membrane current I_m, with the additional reminder that neurons possess varying degrees of cable-like properties.

$$I_m = I_{Cm} + I_{Rm}. \tag{6}$$

That is, the membrane current has both resistive and capacitive components. Substituting for the capacitive current $I_c = C_m dV_m/dt$ gives

$$I_m = C_m \frac{dV_m}{dt} + I_{Rm}, \tag{7}$$

but for a cable, $I_m = \dfrac{1}{r_o + r_i} \dfrac{d^2 V_m}{dx^2}$, where r_o and r_i are extracellular and intracellular resistances respectively, and x is distance.

$$\text{Thus } \frac{1}{r_o + r_i} \frac{d^2 V_m}{dx^2} = C_m \, dV_m/dt + I_{Rm}. \tag{8}$$

Indeed, the soma of the neuron is attached to cable-like axons and dendrites and, in order to achieve good spatial control, the internal potential must be uniform. If it were possible to insert an axial wire throughout the axon and dendrites, such short circuiting would eliminate spatial changes in V_m with distance, that is produce space clamp. As the first term of the Equation 8 is the second spatial derivative of V_m, such a short circuiting axial wire would make the term on the left hand side of Equation 8 equal zero.

$$\text{Thus, } I_m = C_m \, dV_m/dt + I_{Rm}.$$

To eliminate the capacitive current, the voltage clamp is used, because apart from the initial period when the voltage is stepped, voltage clamping makes dV_m/dt equal to zero. Thus in the ideal situation of good spatial control: I_m is equal to I_{Rm}, that is, the voltage-clamped membrane current is equal to the resistive current through membrane ionic channels. In reality, it is not possible to insert an axial wire through axons and dendrites, and the presence of remote processes which are not adequately clamped creates spatial problems. Space clamp problems may be evaluated by determining the dendritic/soma conductance ratio (see Rall in ref 6 and Johnston and Brown's original paper (19)).

If dendrites make little contribution to the overall conductance, then the cell behaves as an isopotential sphere and space clamp is likely to be achieved. Even if space clamping is not possible, membrane currents induced by local application of transmitters to the soma (especially those associated with decreases in membrane conductance) may be adequately clamped. Slowly changing events are also more adequately space clamped (19).

4.2.3 Single electrode voltage clamping (SEVC)

Before discussing the circuit and a typical experiment, it is important to stress that the SEVC is inherently inferior to the two electrode voltage clamp (TEVC, *Figure 7*) with respect to noise and dynamic response. In a practical sense, however, SEVC is superior to TEVC. This is because almost all vertebrate neurons are too small to impale with two electrodes without producing considerable cell damage, additionally many neurons are embedded so deep beneath the microscope fields that they cannot be visualized for impalements with multiple electrodes. Because voltage clamping may be done with a single electrode, SEVC is almost universally used for intracellular microelectrode voltage clamping of vertebrate neurons. The bulk of the ensuing discussion will thus focus on this method.

Circuit and theory

By its very nature, SEVC is limited by the fact that the same electrode is used both for sampling the cell membrane potential (the function of electrode V in *Figure 7*) and for passing current to clamp the membrane potential (the function of the pass I_m electrode in *Figure 7*). To perform these dual functions, the electrode must be rapidly switched between voltage sampling and current passing modes, at frequencies of 1–10 kHz (this rapid switching has lead to the discontinuous SEVC being known colloquially as the 'switch clamp').

A highly simplified circuit diagram of the SEVC is shown in *Figure 9*. The basic scheme is similar to that of the TEVC (*Figure 7*). The main difference is that the output of the preamplifier ($\times 1$) is not fed directly into a clamp amplifier but rather is first delivered into a sample and hold amplifier (SH amp). The output of the clamp amplifier, after receiving the sampled signal and comparing it to the command signal (V_c), is delivered to the switching circuit (Sw), which is under the control of the experimenter. The output of Sw is delivered to a constant current source (CCS); the gain of the CCS (G_{ccs}) determines how precisely the membrane potential approaches V_c (see above for similar description of TEVC, which does not require a CCS to provide optimal gain). Note also the monitoring of the status of the potential across the microelectrode on a separate oscilloscope.

What determines the speed and precision of the SEVC? As illustrated earlier (*Figure 3c*), the microelectrode represents a complex RC circuit, and capacitance in an RC circuit prevents potentials from changing rapidly. Thus the rapid switching is constantly charging and discharging the microelectrode, with the

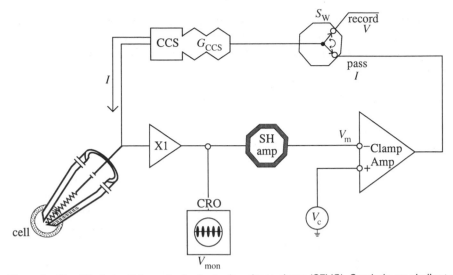

Figure 9. Simplified circuit for a single electrode voltage clamp (SEVC). Symbols are similar to *Figure 8* describing the TEVC. Note the monitoring of the voltage output of the × 1 preamplifier on the oscilloscope (**CRO**) (V_{mon}) prior to the signal being delivered into the sample-and-hold amplifier (SH amp). V_{mon} provides a measure of the voltage drop across the microelectrode; *see Figure 11*. The output of the SH amp, i.e. the sampled V_m is delivered into the Clamp amp, and compared to a command voltage (V_c). The difference between V_m and V_c (recorded in the record *V* mode of the switching circuit, Sw) is fed into a constant current source (CCS) of gain G_{ccs}. High gain in the CCS is required for optimal clamping, much as high voltage clamp gain is required for TEVC. A CCS is used for convenience of circuit design in SEVC.

voltage decline slowed by the presence of distributed capacitance in the circuit. It is only when the voltage drop across the microelectrode has decayed back to baseline that the membrane potential may be clamped with any degree of accuracy. It is thus essential to designate a separate oscilloscope for monitoring the output of the preamplifier (V_{mon}) before it is fed into the next stage (the SH amp) because the V_{mon} signal gives an accurate picture of the charging and discharging of the microelectrode. The maximum frequency at which the clamp may be switched, yet still satisfy the criteria for the microelectrode voltage decaying to zero, is the main determinant of the speed of the clamp. The following is a brief description of what happens during one second of SEVC activity. First the electrode passes current. Next, when the monitored voltage drop across the microelectrode has decayed to baseline, the amplifier is switched by device SW (*Figure 9*) into a sampling mode. The membrane potential is then sampled and held by the SH amp. The next cycle is then begun and the amplifier is switched back into a current passing mode to maintain the sampled V_m at the command level. This cycling occurs at 1– > 10 kHz. The proportion of time current is passed during a cycle is termed the duty cycle (generally set to 30–50%).

From this discussion, it is possible to understand the source of the additional noise in SEVC. First, the recording circuit must have a wide bandwidth to allow the voltage across the microelectrode to settle fully between switching cycles. This requirement increases the noise two-fold over the TEVC. In addition, a digital control signal selects the mode of operation of the SH amp; in 'sample' mode the output of the SH amp follows the input, in the 'hold' mode the value at the moment of switching is held. This digital control of sampling introduces an additional doubling of the noise due to aliasing (see below for discussion).

An actual experiment

The following treatment describes an actual experimental protocol for the use of the SEVC. As the SEVC experiment begins in the bridge mode, this protocol will begin with a review of the bridge method for current clamp. The experiment then moves into the discontinuous current clamp mode (in which the sample and hold amplifier is employed under current clamp) and then to SEVC.

Protocol 1. Single electrode voltage clamping (SEVC)

Begin in the bridge mode, passing current pulses and measuring potential changes (see *Figure 5*).

1. With the electrode extracellularly, use capacitance neutralization to improve the frequency response of the unbalanced potential (*Figure 5a*). Do not balance the bridge until negative capacitance has first been employed (spurious bridge balance might result, see ref. 4).

2. Impale the cell, obtaining a stable V_m before proceeding. Perform experiments in current clamp as described above, ascertaining that bridge balance is maintained throughout, or proceed with the following protocol.

3. Switch into a discontinuous current clamp (DCC). This provides a convenient way to use capacitance neutralization in conjunction with the SH amp.

4. Again attempt to obtain the maximum rise time on the V_{mon} trace without sending the amplifier into oscillation. This is important as the better the capacitance neutralization, the faster the settling time of the clamp. *Figure 10* shows three traces: **a**, underutilized; **b**, optimum; and **c**, overutilized capacitance neutralization (on the voltage output, poorly-adjusted capacitance neutralization resembles improper bridge balance (not shown)). DCC is much noisier than a conventional bridge circuit but allows V_m to be recorded without an IR drop across the microelectrode. DCC is thus less susceptible to errors introduced by small changes in electrode resistance and is, on occasion, used in lieu of a bridge circuit for measuring passive membrane properties (see legend to *Figure 5*).

5. Switch into SEVC mode. Clamp the membrane at the resting potential and generate command voltage pulses (for example 10 mV), much like current pulses under current clamp. Monitor the voltage (V_{mon}) across the microelectrode on a separate oscilloscope. This is the crucial part of the SEVC procedure. What one observes is a series of variable bell-shaped

Protocol 1. *Continued*

voltage traces, the variability due to inherent sweep-to-sweep variations in the sampled voltage noise (*Figure 11b*).

6. Increase the switching frequency gradually to the maximum possible while still allowing the voltage across the microelectrode to decay totally. The sampling rate (cycle rate) is chosen so that at least 10 cycles occur within the time constant of the membrane, thus allowing C_m to smooth the membrane voltage response. For example, if τ_m is 50 ms, then the sampling rate would be set higher than 2 kHz to allow for 10 cycles to occur in the 10 ms period. The higher the sampling rate, the lower the noise level (see Finkel and Redman in refs 6, 9).

7. Increase the clamp gain setting to just below the maximum attainable level without causing instability (ringing or overshooting in the voltage pulses).

8. If *and only if* the microelectrode voltage obviously deviates from a simple exponential time course and has evident fast and slow phases of decay, use anti-aliasing filtering to filter the slower phase of decay. *Explanation*: aliasing occurs when the rules of data sampling are not followed properly. The Nyquist sampling theorem requires that a signal must be sampled at a minimum of twice its highest frequency component. Thus a signal with frequency components extending from 0–1 kHz should be sampled at 2 kHz (i.e. every 0.5 ms). If the sampling rate is too low, high frequency noise is translocated to lower frequencies. This is termed *aliasing* and constitutes a nuisance to faithful SEVC. Aliasing is akin to watching the wheels of a stagecoach in an old western film. In this analogy, the wheels are actually turning at high frequency but, because the camera undersamples the frequency of revolution, the wheels appear to be turning very slowly. Noise associated with aliasing appears in the slower phase of decay of the microelectrode voltage. This slow decay may be due to the distributed nature of C_t (*Figure 3c*), the redistribution of ions in the tip of the electrode, or other less obvious causes. The anti-aliasing filter improves the input signal before being fed into SH amp, reducing the noise in the slow phase of decay without altering its time course (the time course of this slow phase is rate-limiting as far as settling time of the clamp). Unfortunately, such filtering may also reduce the maximum sampling rate. The experimenter thus has to trade-off noise elimination against the optimal dynamic response of the system.

9. Adjust the phase but use it sparingly. *Explanation*: the voltage clamp amplifier requires a 90° phase shift in the circuit to perform its negative feedback function optimally. Such a phase shift is supplied, ideally, by the capacitance of the membrane. Phase lead increases the high frequency gain and can sharpen the step response of the system, but may also increase noise and induce oscillations. Phase lag does the opposite and slows the response but also reduces noise. Practically, a certain amount of phase lag may be necessary to allow for optimum clamp gain to be achieved. Introduce phase shifts if *both* current and voltage responses may be improved by such

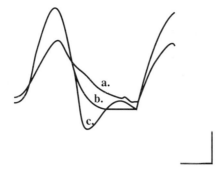

Figure 10. Capacitance neutralization under discontinuous current clamp. Records show the output of the × 1 preamp (V_{mon}) of *Figure 9* under current clamp using the SH amp. (a) shows underutilized; (b) appropriate, and (c) overutilized capacitance neutralization. Note the decay reaches its final value rapidly in (b). Calibration bars: vertical, 40 mV; horizontal, 10 μs.

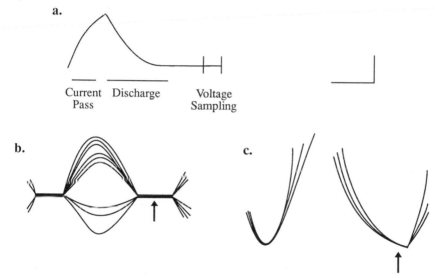

Figure 11. Monitoring of the voltage signal across the microelectrode during a SEVC experiment. In the idealized record, **a** current is passed, the voltage across the micro-electrode is allowed to discharge and the voltage is sampled during the period of a flat baseline; **b** shows acceptable clamp in which the voltage drop across the microelectrode has decayed to baseline before sampling (arrow); **c** shows false clamp in which signal across the electrode is still decaying at the time of sampling. Calibration bars: vertical, 20 mV; horizontal, 50 μs.

changes. Caution must be made in using the phase adjustment. This is because excessive phase adjustments can cause an IR voltage drop across the microelectrode to appear spuriously as a change in measured membrane potential, thus producing a false clamp. If the electrode voltage monitored as V_{mon} does not decay back to baseline, then a false clamp may be indicated.

Protocol 1. *Continued*

10. Assess again the status of the voltage across the microelectrode (V_{mon}). It cannot be overstated that the experimenter must ascertain that fluctuating transients all decay to baseline well before the measurements of membrane voltage are made (see *Figure 11b*, arrow) as even the smallest residual deflection prior to sampling will cause serious measurement errors. *Figure 11c* shows the improper clamp situation. These assessments must be made at the beginning and *during the entire time course of the experiment* as changes in microelectrode resistance during prolonged current passing can alter the clamp settling parameters and produce a false clamp. It might even be prudent, before beginning the actual experiment, to pass very large steady hyperpolarizing currents (for example 1 nA) across the electrode once the optimal clamp parameters are established at the resting potential to examine how well the membrane remains clamped during the passage of current.

11. Begin the actual experiment.

In the experiment shown in *Figure 12a*, 600 nM ATP is applied to a cell voltage clamped at the resting membrane potential. Note that ATP produces an inward current (downward) associated with a conductance increase, i.e. an increase in the size of the transient currents produced in response to the 10 mV depolarization (for a constant voltage transient, by Ohm's law an increase in the conductance is reflected as an increase in the transient current).

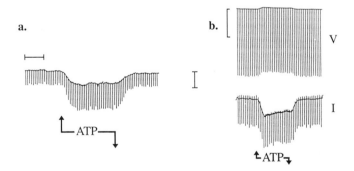

Figure 12. ATP-induced inward currents under SEVC. (a) Shows ATP (600 nM) produces large inward currents and an increase in membrane conductance (decreased resistance) as is evidenced by an increase in the size of the current pulses (spikes) recorded in response to constant voltage pulses (not shown). (b) Shows both current (*I*) and voltage (*V*) traces from another experiment to illustrate an inadequate clamp. Note that in (b), the voltage trace, rather than being unchanged in the presence of ATP (which would indicate adequate voltage clamping) slowly depolarizes just as it did in the unclamped condition (for example *Figure 7a*). Horizontal calibration, 20 s; vertical cal. in middle, 300 pA for **a** and lower trace in (b). Vertical cal. to left of V trace, 15 mV (unpublished data).

Thus conductance increases are reflected as decreases in electrotonic potentials in unclamped neurons and as increases in transient current responses under voltage clamp. The advantage of the SEVC experiments is that the conductance change is not due to voltage-gated ion channels (as the membrane is voltage clamped) but is due to the action of ATP to open its own fast ion channel with a reversal potential in the depolarizing direction.

Figure 12b shows an experiment in which the membrane is not adequately clamped. Note that while the current (*I*, lower trace) appears as before (*Figure 12a*), the voltage (*V*, upper trace) slowly rises in the presence of ATP, much as it did in the non-voltage clamp experiment (*Figure 6a*). This occurs even in the presence of sufficient tetrodotoxin to block currents through the voltage-gated Na^+ channels associated with the action potential. It is apparent that the microelectrode cannot supply sufficient current to clamp the membrane potential in the face of the large conductance increase produced by ATP (as the voltage trace should be unchanged by ATP if the SEVC were adequate). Indeed, in association with large conductance increases (i.e. resistance decreases), the time constant of the membrane changes declines radically, thus altering the properties of the switch clamp. The clamp may thus be well set at rest but behave poorly during a large conductance increase. SEVC works optimally for modest conductance increases in cells with little dendritic influence or under conditions where a substance increases the resistance of the membrane.

4.2.4 Two electrode voltage clamp (TEVC)

A simplified circuit for the TEVC has already been presented in *Figure 7*. We, and others, have used the TEVC in studies of the action of ACh on adult skeletal muscle (20), on the expression of various ionic channels and receptors in frog oocytes (21), and to study the properties of larger invertebrate neurons (22). However, either due to the delicacy of the neuron or to the lack of microscopic visibility, TEVC is only seldom used in studies of 'uncultured' neurons of the vertebrate nervous system. A notable exception is the work of Belluzzi and Sacchi (23) who have performed a series of studies on rat sympathetic neurons *in vitro* using TEVC. *Figure 13* summarizes some of their results on the voltage-gated Na^+ currents involved in the upstroke of the action potential. *Figure 13a* shows the effects of two intracellular electrodes in the unclamped neuron. Note both electrodes record action potentials in response to depolarizing current through the microelectrode (part 1) and to synaptic activity associated with the esp (part 2). *Figure 13* shows the Na^+ currents under voltage clamp, the neuron being held at $-70\,mV$ and then stepped to the indicated membrane potential. Averaged current–voltage relationships for Na^+ channels are shown in *Figure 13c*.

If the investigator is fortunate enough to discover a vertebrate neuron amenable experimentally to TEVC, a few technical comments might be in order. Firstly, the need for high clamp gain, capacitance neutralization, and phase control adjustments discussed above for SEVC are also important aspects of

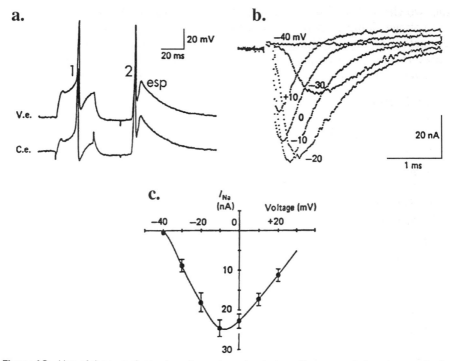

Figure 13. Use of the two electrode voltage clamp to study sodium currents in rat sympathetic neurons. (a) Responses to current injection through one electrode with both electrodes recording only voltage. V.e. is destined to be the voltage recording electrode in the voltage clamp condition and C.e. passes current in both conditions. Action potential at (1) is due to depolarizing current supplied by C.e. Action potential at (2) is due to presynaptic nerve stimulation (note esps). (b) Na$^+$ currents under TEVC. Currents were evoked by voltage steps from -70 mV to the indicated potentials. (c) Averaged I_{Na}–Vm relationships for 10 cells. Reprinted with permission from ref. 23.

TEVC. Anti-alias filtering is not used with TEVC, however, as it tends to make the voltage-recording electrode sluggish. There is one major technical problem in TEVC, namely eliminating the coupling capacitance between the two electrodes due to the air and bathing fluid present between the electrodes. Such capacitance, even if as low as 0.01 pF may affect the high frequency response of the amplifier and lead to oscillations. The solution to this problem is to provide some form of grounded shielding between the two electrodes, such as wrapping current electrodes in aluminium foil (very near to the tip and insulated from the bathing fluid), or using a grounded helically-arranged wire shield as part of the microelectrode holder (*Figure 7*). Other TEV clampers coat the current electrode with conducting paint. Series resistance problems emanating from a number of sources may become apparent when large conductance increases occur. Series resistance may be due to membrane infoldings or to the coupling resistance in the region of two electrodes (high current density in this region causes a voltage drop across the fluid coupling the two electrodes). Voltage drops across series resistances are interpreted by the voltage clamp as part of the measured Vm. It may thus be necessary to estimate the value of the series resistance and subtract the IR drop across the series resistance electronically. For further details of the advantages and disadvantages of these methods, see Chapter 4 of ref. 6, and pages 20–2 of ref. 7.

4.2.5 Summary of voltage clamping

From the above discussion, it appears that if the situation is at all amenable to stable impalements without damaging the cells using two electrodes, then TEVC should be employed. TEVC is inherently lower in noise, and allows faster response time and better clamping than SEVC. As the vast majority of neurons do not allow the experimenter this luxury, SEVC is the most widely used tool to voltage clamp vertebrate neurons *in vitro* using intracellular microelectrodes. On the positive side, the SEVC:

(a) does not have the problem of coupling between electrodes;

(b) is a very convenient way to study reversal potentials for slowly changing membrane currents;

(c) is free of problems due to voltage drop across the microelectrode series resistance as this is not clamped with switch clamping; and

(d) may be used to advantage with patch pipettes. These low resistance electrodes (see Chapter 3) allow switching frequencies > 10 kHz to be employed and thus reduce noise levels. Series resistance problems attributable to the patch electrode resistance may also be eliminated by this method. Indeed in large cells that may be impaled with low resistance electrodes, using a very fast switching rate and exploiting its lower associated noise, the advantages of the SEVC may approach those of TEVC.

With TEVC, series resistance problems may appear and often constitute a major impediment to studies with *continuous* single electrode voltage clamping, one variant of which is the whole cell patch clamp. Patch clamping is discussed in Chapter 3 of this volume.

5. Final comments

It is hoped that within this limited space, the reader has obtained some useful information on intracellular recording methods from neurons. Despite attempts to generalize, each neuron possesses its own optimal method for experimental scrutiny and the electrophysiologist must be aware of differences in neuronal size and dendritic arborization and exploit these differences for successful studies of membrane behaviour.

Acknowledgements

I wish to thank Dr S. Vanner for his collaboration on the experiments illustrated in *Figures 6a* and *12*, and Dr Vanner and Dr John P. Williams for valuable comments on the manuscript. I am also indebted to Ms June Shiigi for her illustrative assistance. This work was supported by a research grant from the US Public Health Service (NS 12782).

References

1. Hoyle, G. (1983). *Trends Neurosci.*, **6**, 163.
2. Hogg, B. M., Goss, C. M., and Cole, K. S. (1934). *Proc. Soc. Exp. Biol. Med.*, **32**, 304.
3. Ling, G. and Gerard, R. W. (1949). *J. Cell. Comp. Physiol.*, **34**, 383.
4. Purves, R. D. (1981). *Microelectrode methods for intracellular recording and iontophoresis.* Academic Press, London.
5. Geddes, L. A. (1972). *Electrodes and the measurement of bioelectric potentials.* Wiley, New York.
6. Smith, T. G. Jr., Lecar, H., Redman, S. J., and Gage, P. W. (1985). *Voltage and patch clamping with microelectrodes.* Am. Physiol. Soc. Waverly Press, Baltimore, MD.
7. Standen, N. B., Gray, P. T. A., and Whittaker, M. J. (1987). *Microelectrode techniques: the Plymouth workshop handbook.* Company of Biologists, Ltd. Cambridge, England.
8. Wilson, W. A. and Goldner, M. M. (1975). *J Neurobiol.*, **6**, 411.
9. Finkel, A. S. and Redman, S. (1984). *J. Neurosci. Meth.*, **11**, 101.
10. Silinsky, E. M. (1987). In In vitro *methods for studying secretion.* (ed. A. M. Poisner and J. M. Trifaro), pp. 255–271. Elsevier, Amsterdam.
11. Woodbury, J. and Crill, J. (1951). *Am. J. Physiol.*, **164**, 307.
12. Tasaki, K., Tsukuhara, Y., Ito, S., Wayner, M. J., and Wu, W. Y. (1968). *Physiol. Behav.*, **9**, 120.
13. Blackman, J. G., Ginsborg, B. L., and Ray, C. (1963). *J. Physiol.*, **167**, 355.
14. Hirst, G. D. S. and Silinsky, E. M. (1975). *J. Physiol.*, **251**, 817.
15. Thomas, R. (1972). *J. Physiol.*, **220**, 58.
16. Adrian, R. H. (1956). *J. Physiol.*, **133**, 631.
17. Mihara, S., North, R. A., and Surprenant, A. M. (1987). *J Physiol.*, **390**, 335.
18. Ginsborg, B. L., House, C. R., and Silinsky, E. M. (1974). *J. Physiol.*, **251**, 817.
19. Johnston, D. and Brown, T. H. (1983). *J. Neurophysiol.*, **50**, 464.
20. Silinsky, E. M. and Vogel, S. M. (1987). *J. Physiol.*, **390**, 33.
21. Christie, M. J., North, R. A., Osborne, P. B., Douglass, J., and Adelman, J. P. (1990). *Neuron*, **2**, 405.
22. Gorman, A. L. F. and Mirolli, M. (1972). *J. Physiol.*, **227**, 35.
23. Belluzzi, O. and Sacchi, O. (1986). *J. Physiol.*, **380**, 275.
24. Silinsky, E. M. and Ginsborg, B. L. (1983). *Nature*, **305**, 327.

Appendix 1: Distinguishing between presynaptic and post-synaptic effects of modulatory substances

There are several ways to determine whether the depression of synaptic transmission illustrated in *Figure 1* is due to a presynaptic or post-synaptic effect of the exogenous inhibitor.

(a) The best way, if possible, is to measure small, miniature esps that occur in the absence of nerve stimulation; such measurements allow for accurate moment-to-moment assessments of physiologically-functional transmitter release. Specifically, the ratio of the mean esp evoked by nerve stimulation to the spontaneous miniature esp indicates the number of transmitter quanta released synchronously by a nerve impulse.

(i) If the ratio of the mean evoked esp to mean miniature esp is decreased by X, then X is exerting a presynaptic effect. In the experiment, shown in *Figure 1*, X is ATP, a presynaptic inhibitor of ACh release as it reduces the ratio of evoked to spontaneous esp amplitude (24). Less ideally, the smallest class in the evoked release histogram may be used in place of the miniature.

(ii) Presynaptic effects may also be reflected as a change in the frequency of occurrence of miniature potentials but this is not always the case.

(b) In situations where miniature potentials are not clearly distinguishable from the noise, other methods are required.

(i) For example, the neurotransmitter may be applied exogenously by fast flow delivery, by local iontophoresis (using electric current to expel charged drug from the drug-containing microelectrode), or by pressure-ejection (using a 'picospritzer' to puff the neurotransmitter from microelectrodes positioned at the surface of the neuron). If drug X reduces the response to the exogenously applied transmitter, then a *post-synaptic* locus is likely to be responsible for the synaptic inhibition. It should also be noted that a decrease in the *size* of the miniatures is most frequently due to a reduction in sensitivity to neurotransmitters, although presynaptic effects may occur if the transmitter stores have been subject to depletion by persistent presynaptic stimulation.

(ii) Another way to assess presynaptic depression, if no miniature potentials are detectable, is to compare the effects of a series of closely spaced stimuli on the pattern of esps in the presence and absence of X. If X *decreases* the size of the esp to *individual* widely spaced stimuli yet *increases* the level of *facilitation* of subsequent esps to brief high frequency stimulation, then such a result is suggestive of presynaptic inhibition. This is because, when a smaller esp is due to smaller amounts of transmitter release, there is less depletion of transmitter with the small esp as compared to the larger control esp. Thus, with less depletion of the first esp, depression of subsequent esps in response to rapid repetitive nerve stimulation is less than in the control situation, allowing increased facilitation to occur as it is less encumbered by the concomitant depression.

Generally, experiments on synaptic potentials are made using magnesium ions to reduce the control esp amplitude below threshold (by a presynaptic depressant action) so that the effects of modulatory substances may be more accurately quantified. For specific details of averaging methods and equations used to correct for various membrane non-linearities in the quantitation of synaptic communication see ref. 10.

Appendix 2: Equipment and suppliers

This section provides additional details on choices of instrumentation and other appurtenances necessary for intracellular recording. An additional focus of this section relates to alternatives between purchasing new equipment or relocating unused equipment from vacant laboratories or storerooms into the investigator's laboratory.

Capillary glass tubing for microelectrodes—best purchased from WP instruments or an equivalent supplier. Be certain to purchase tubing with glass fibres affixed inside.

Microelectrode pullers—Sutter Instruments the Brown–Flaming puller includes timed cooling jets of gas and microprocessor controlled settings of temperature, velocity, and pull strength. Alternatively, older Narishige models are especially recommended if electrodes less than 40 MΩ suffice for stable, damage-free recordings.

Preamplifiers and voltage clamp amplifiers—for purchasers, if a single piece of equipment is required, one of the Axoclamp series (Axon Instruments) is highly recommended. These are superb instruments that can perform intracellular recording, current clamp (both continuous and discontinuous), SEVC, TEVC, whole cell patch clamp, and can even be used as an iontophoresis device for passing small amounts of charged drugs or neurotransmitters from microelectrodes on to receptive surface of cells. Alternatively, WPI or Dagan preamplifiers, and single and two electrode voltage clamp amplifiers may often be found on storage shelves. While these amplifiers do not have buzz switches like the Axoclamp amplifiers, rapid turning of the capacitance neutralization knob may be used in lieu of the buzz control.

Oscilloscopes—the Tektronix 500 tube series continue to be excellent even after 30 years of use. Solid state equivalents and digital oscilloscopes may also be purchased from Tektronix if desired. Try to obtain an oscilloscope with an output of one volt per division regardless of the gain of the amplifier. Such outputs are useful for feeding pen recorders and tape recorders.

Computer-assisted data acquisition and analysis—the most widely employed system may be purchased from Axon Instruments. It includes the A/D D/A converter (TL-1 interface and Labmaster board) and pCLAMP software for data capture and analysis. A discriminator may be required for capturing spontaneous events. This system is IBM-PC-based and is excellent for studying membrane currents controlled by the investigator and for digitizing esps and evoked synaptic currents. Additional software or a dedicated programmer is required to perform detailed analysis of transmitter release using pCLAMP, however. Excellent IBM-based software for synaptic events has been written by Dr John Dempster in the Department of Physiology at Strathclyde University, Glasgow G1 1XW, UK, and is in use in Canada and elsewhere in the UK. Hewlett–Packard XY plotters and Gould pen recorders in conjunction with hand-held scanners are popular for producing hard copy.

Pen recorders—if an excellent recorder is desired, it is best to purchase one from Gould Instruments. A computerized recorder (Axotape from Axon Instruments) is a viable alternative. DC calibrators (for example WP Instruments) placed in series with the reference electrode are useful for offsetting resting potentials and preventing recorder pens from being offscale.

Magnetic tape recorders—VCRs (*Figure 1*) are generally employed (INDEC, *Figure 1*) although FM tape recorders (Hewlett–Packard, Racal) provide excellent flexibility as the signal may be played back at different speeds for producing hard copy.

Antivibration mountings—these vary from balance tables placed in beach sand with magnetic manhole covers atop tennis balls or bicycle tyres, to true air tables (TMC Manufacturing or Newport).

Microscopes—these vary depending upon the needs of the investigator. Anything from a basic dissecting 'scope to a superior dissecting 'scope to inverted microscopes equipped with phase contrast, Hoffman or Nomarski optics (Leitz, Carl Zeiss, Nikon) are possible.

3

Patch clamp methods for single channel and whole cell recording

N. B. STANDEN and P. R. STANFIELD

1. Introduction

The patch clamp technique was first described by Neher and Sakmann (1) and further developed by Hamill *et al*. (2). It has led to major and widespread advances in the electrophysiological analysis of the behaviour of cell membranes for two main reasons. First, it has allowed the resolution of currents through single ion channel proteins in native cell membranes. This has led to detailed studies of their permeability and kinetic properties, pharmacology, and modulation, as well as providing powerful new criteria for the classification of channels, leading to the identification of several new types. Secondly, patch clamp has increased the range of cells suitable for electrophysiology to include many (especially mammalian) types too small or fragile for previous techniques.

Patch clamp uses an electrolyte-filled glass micropipette electrode, with a tip diameter typically 1 μm or less. The tip is fire-polished so that the pipette rim forms a tight seal with the cell membrane rather than penetrating the cell. This electrically isolates a small patch of cell membrane and allows ion flow through ion channels in the patch to be measured as electrical current. The current through a single ion channel is usually in the range 0.1–10 pA, so that a sensitive amplifier and careful attention to reduction of extraneous electrical noise are needed to study single channel currents. Variants of patch clamp allow patches to be detached (excised) from the cell, so that either their cytoplasmic face (inside-out patch) or extracellular face (outside-out patch) is exposed to the bath solution. For recording whole cell currents the membrane patch is destroyed after seal formation, either by suction or electrically, or it may be permeabilized with the pore-forming antibiotic nystatin. This gives electrical access to the inside of the cell, allowing currents through the whole of the cell membrane to be measured under voltage clamp, or the membrane potential to be measured under current clamp (see Chapter 2).

The basic patch clamp circuit is shown in *Figure 1*. The amplifier illustrated forms the headstage of commercial patch clamp amplifiers and acts as a current-to-voltage converter. The amplifier makes the voltage in the patch pipette

Patch clamp recording

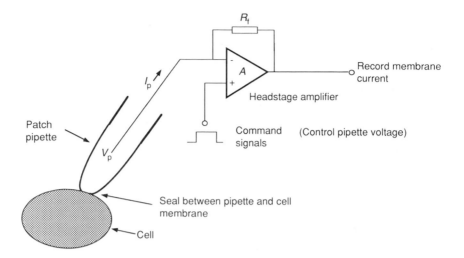

Figure 1. Basic patch clamp recording. The amplifier A is connected in negative feedback mode as a current to voltage converter. Its output will be given by $(-I_pR_f + V_c)$, where V_c is the voltage of the command signal. The pipette potential V_p is clamped to V_c. Commercial patch clamp amplifiers provide many additions to this basic circuit.

(V_p) follow the command voltage, and current is measured as the voltage drop across the feedback resistor R_f. For single channel recording R_f needs to be large (10–100 GΩ) to minimize noise. This limits the maximum current, however, so that a lower R_f value (0.5–5 GΩ) is often needed for recording whole cell currents. For this reason most commercial amplifiers have switchable R_f values. A new development for single channel recording is the integrating headstage, in which a feedback capacitor is used in place of R_f. This gives lower noise at the expense of a brief interruption in recording from time to time when the capacitor is discharged.

A high seal resistance between pipette and cell membrane reduces the current that flows through the seal when V_p is not zero. This is normally seen as a steady holding or leakage current, and is sometimes compensated for with analog circuitry. A high seal resistance also minimizes noise, since the variance of the current noise in a resistor varies inversely with its resistance R, being given by:

$$\sigma_I^2 = 4kTB/R \tag{1}$$

where k is Boltzmann's constant, T is absolute temperature and B the recording bandwidth (usually set by a low-pass filter). Thus a tight seal (~ 2–100 GΩ) is usually essential for single channel recording, where the resistance of the membrane patch is normally much higher than that of the seal, while relatively tight seals also make whole cell recording easier.

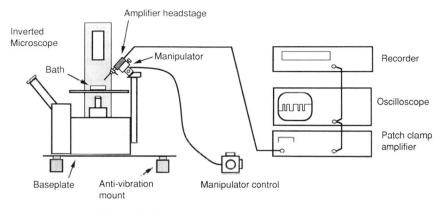

Figure 2. Components of a patch clamp set-up.

Many further details of the theory of patch clamp recording and of patch clamp amplifiers may be found in refs 3–6. This chapter will concentrate on the practical aspects of patch clamp recording.

2. Basic apparatus

A certain minimum amount of equipment is needed for patch clamp. Some items are quite expensive, so that the capital cost of a patch clamp set-up is relatively high; recurrent costs, however, are usually low. Components of a typical recording set-up are shown in *Figure 2*. The basic requirements are as follows:

- microscope
- micromanipulator
- baseplate or anti-vibration table
- patch clamp amplifier
- oscilloscope and other display and recording equipment

In addition, the following are needed to make electrodes. They are discussed in Section 5:

- electrode puller
- microforge for fire polishing

2.1 Microscopes

A microscope with phase contrast, or differential interference contrast (Nomarski or Hoffman) is used to view the preparation and to position the electrode. Inverted microscopes are commonly used, though non-inverted ones are sometimes preferable when the preparation or tissue is such that it is hard to see the approach of the electrode through it. An inverted microscope should

61

have a long-working distance condenser, while a non-inverted microscope will need a long-working distance objective. Overall magnifications of $\times 200$–600 are typically used to make seals, while a low power objective can help in locating cells and in initial electrode positioning. Further requirements are that the microscope should be mechanically stable and have a stage that does not move when focusing. Commonly used microscopes are those made by Nikon (Diaphot or TMS), Carl Zeiss (ID02), and Leitz.

2.2 Micromanipulators

The micromanipulator should be capable of fine movement in three axes, and give good isolation of the experimenter's hand movements from the preparation. Two types in which the controller is remote from the manipulator head, and so may be mounted off the baseplate, are available:

● Hydraulic manipulators, filled with oil or water. Some models may drift if the temperature changes. Available from Narashige.

● Electrical manipulators. Motorized types, drive by DC or stepping motors, are made by Märzhauser. Piezoelectric manipulators are also available, for example from Burleigh.

Both hydraulic and motorized manipulators give about 1 cm of remotely controlled movement in each axis, while piezoelectric types give much less ($< 100\,\mu m$). All can be bought with integral (non-remote) coarse movements giving a greater range for initial electrode positioning.

Alternatively, high-quality, non-remote manipulators, such as Huxley–Goodfellow or Leitz types, may be used. To minimize vibration, manipulators are usually mounted on the microscope stage or on the body of the microscope. The headstage of the patch clamp amplifier is mounted on the manipulator (usually with double-sided tape) and has a socket for the electrode holder.

2.3 Baseplates or anti-vibration tables

To reduce vibration, the microscope, manipulator, and headstage should be mounted on a steel baseplate 1–2.5 cm thick. Commercial pneumatic anti-vibration tables are widely used, for example those made by Ealing or Wentworth. These usually include both baseplate and legs. Alternatively, a steel plate may be placed on inflated inner tubes (small bicycle or scooter) on a solid table or bench. The extent to which vibration is a problem depends on building design. In some buildings, the best solution is to use the ground floor.

2.4 Patch amplifiers

Several good patch clamp amplifiers are available. In addition to the basic headstage circuit of *Figure 1*, these amplifiers offer many other features. Additional gain stages give final amplification of up to 1000 mV/pA, and high frequency boost circuits improve bandwidth. Most amplifiers also have the following:

- junction potential offset
- tracking circuit ('search') to zero current during seal formation
- voltage clamp and current clamp mode
- potentiometer to set holding potential
- compensation for electrode and cell capacitance
- compensation for series resistance in whole cell recording
- meters giving readouts of pipette voltage and current
- low-pass filter for the output current signal

Some amplifiers also provide leakage subtraction, a test pulse for seal formation, and an audio tone which varies with pipette current. The List-electronic EPC-7 and Axon Instruments Axopatch 1 series amplifiers are very widely used; others are made by Dagan and Bio-Logic (RK 300). The Dagan 3900 and Axopatch 200 have integrating headstages for single channel recording, but switch to a feedback resistor for whole cell recording, while the List EPC-9 is controlled by an integral microcomputer. Patch clamp amplifiers are usually supplied complete with electrode holders.

2.5 Display and recording equipment

The basic item of equipment for displaying the current or voltage signal from the patch clamp amplifier is an oscilloscope, which may be analog or digital. Storage is helpful in either case. The output may also be recorded on a chart recorder, though limited bandwidth may lose high frequency events. Videotape recorders with a modified pulse code modulator (PCM) (7) are an economical means of storing large amounts of data, and modified digital audio tape recorders are available for the same purpose. A low-pass 8-pole Bessel filter is very useful both for filtering data for display as it is collected, and for filtering when data is digitized for analysis. It is also possible to use a microcomputer system with A/D (analog to digital) and D/A converters to sample and display patch clamp recordings and to provide command voltages to the patch amplifier, as well as for later analysis of recordings. Most existing systems are based on PC computers, and systems and software packages are available, for example, from Axon Instruments, Cambridge Electronic Design, and Indec.

3. Reducing electrical noise

Single channel currents are very small, but even for recording from whole cells it is important to reduce electrical noise as much as possible. The noise is of two types; external interference from electrical sources in the vicinity of the recording set-up, and intrinsic noise in the seal and recording system itself.

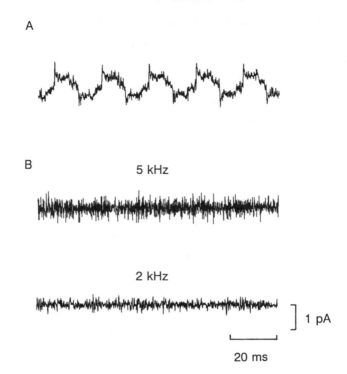

Figure 3. Noise in patch clamp recording. The traces show records of the current from a gigaohm seal, recorded using a patch clamp amplifier. A. An example of interference noise. The deflections in the record are synchronous with the mains frequency, and usually occur at this frequency or a multiple of it. 2 kHz filter. B. Intrinsic noise. This noise is essentially random. The upper trace is filtered at 5 kHz (8-pole low-pass, Bessel), and the lower trace at 2 kHz.

3.1 Noise from interference

Interference results when the recording set-up picks up signals from mains cables, power supplies, lights, and so on, which will be present in any building. These signals are usually synchronous with the mains frequency, and can, therefore, be most easily seen if the oscilloscope used to display the output of the patch clamp amplifier is triggered from the mains frequency (*Figure 3A*). Most oscilloscopes have a button or switch on their trigger panel enabling you to do this. Removal of interference noise is achieved by shielding the most sensitive parts of the recording set-up, and by careful earthing of the patch clamp equipment.

3.1.1 Shielding

Overall shielding is normally provided by a Faraday cage (an earthed wire mesh, or sometimes sheet metal, cage surrounding the baseplate, microscope, preparation, and headstage of the patch clamp amplifier). The cage should be mounted so that it does not transmit vibration to the baseplate, for example by standing it on the floor independently of the air table. The front of the cage is usually either hinged or made of a curtain of hanging chains, so that the experimenter has access to the preparation. For most set-ups, much shielding will also be provided by the microscope. This should be earthed, and it is also important to check for electrical connection between different parts of the microscope. This can be done with a resistance meter. If, for example, stage and body are not connected, such a connection must be provided or they must be earthed separately. Other equipment within the cage, for example micro-manipulators, should also be earthed. One exception is the patch clamp amplifier headstage case, which is often driven to a potential different from earth, and should, therefore, be mounted on the micromanipulator in such a way that it does not make electrical contact. Interference often arises from sources of electrical power which come into the cage; for example microscope light sources or power for electrical manipulators. Particular attention should be paid to shielding these (see also Chapter 1). If problems persist, it is often possible to arrange DC power supplies for such equipment.

3.1.2 Earthing

The quality and pattern of earthing is critical for removing interference noise from small signals. The basic principle is to earth equipment; microscope, baseplate, cage, and so on to a common earth point using low resistance leads. Patch clamp amplifiers provide a high quality earth terminal on the amplifier case to which this common earth point can be connected. (The earth terminal on the amplifier headstage itself is normally used only to connect the bath electrode; see Section 6.) It is important to avoid 'earth loops' which can act as aerials picking up interference. An earth loop occurs when two pieces of equipment that are both connected to the common earth point also have their earths connected together by another route. If the earthing pattern is thought of as a tree with the common earth point as the trunk, it is important not to connect between the branches. After following these basic rules, the removal of interference noise is usually a process of trial and error, involving experiments with slightly different patterns of earthing. Some further specific details are often given in the manuals of patch clamp amplifiers.

3.2 Intrinsic noise

This is thermal noise arising from the seal, patch membrane, pipette and holder, and patch clamp amplifier itself. It is random, appearing as a uniform broadening of the displayed current record (*Figure 3B*), and thus not syn-chronous with the mains frequency. Its absolute level is of most concern when recording from single channels. Most patch clamp amplifiers allow you to measure and display the root mean square value of random noise. In an ideal recording situation pipette, seal, and amplifier contribute about equally to the random noise (3, 6). The main ways of reducing this noise as far as possible are as follows:

- use hard glass for patch pipettes (Section 5)
- coat pipettes with Sylgard (Section 5)
- use a pipette holder made from a high-dielectric, non-wettable plastic like Teflon
- keep the pipette holder clean and dry
- keep the length of the pipette dipped into the bath as short as possible

3.2.1 Filtering

The level of noise finally sets the resolution of patch clamp recording, especially for single channel work. Low-pass filtering is used to improve the signal:noise ratio. Although some filtering is provided by the patch clamp amplifier, often an additional external filter is also used. The variance of the noise decreases in proportion to the bandwidth (see Section 1), but the price for the improvement in signal:noise is loss of temporal information. The theory of filtering and criteria for setting the bandwidth are discussed fully in references 8 and 9.

4. Preparations suitable for patch clamping

Given a modicum of skill on the part of the experimenter, success in carrying out single channel recording seems to depend on the preparation. In general the choices are:

- adult or young animal cells, which usually have at least been cleaned enzymatically and which may have been dissociated
- membrane vesicles formed from adult cells
- primary cell cultures
- cell lines grown up in tissue culture.

Most laboratories that undertake patch clamp work end up using cell culture at some stage and many find it necessary to develop their own cell culture facility. An account of cell culture methods is given in ref. 10, while accounts of the culture of nerve cells may be found in refs 11 and 12.

Problems with adult cells include the likelihood that channels will be present at very high density (for example sodium channels are present at densities $> 100\,\mu m^{-2}$). Problems with primary cultures include uncertainties about the expression of ion channels or the receptors for neurotransmitters which modulate channel function. There may be uncertainty with some cell lines as to how fully they represent the normal situation. In addition, some may be particularly difficult to seal on, though others may be among the easiest preparations to keep in culture and to use in recording.

4.1 Adult cells

These should usually be dissociated. Most people dissociate cells using either collagenase, protease, trypsin, or papain. Neurons may be freshly dissociated from brain using papain or trypsin in Ca^{2+} and Mg^{2+}-free medium. A suitable method is given in ref. 13. In their method, the brain is sliced with a tissue chopper, or cut up with small scissors. The pieces of brain are subjected to digestion with trypsin in oxygenated saline. The cells are dissociated by trituration with fire-polished Pasteur pipettes with a small (0.2–0.5 mm) tip diameter in culture medium. It is imperative to wash off culture medium, replacing it with saline, before attempting patch recording since the presence of protein makes sealing very difficult.

Methods for isolating cardiac myocytes are given in ref. 14, and methods for smooth muscle cells by, for example, refs 15, 16, while a more general discussion of dissociation methods may be found in ref. 17.

New methods of working with neurons include the use of very thin brain slices (18), made with a vibrating microtome (such as the Lancer Vibratome, A. R. Horwell) used to cut slices 100 µm thick. The brain must be kept cold (4 °C) during slicing, preferably using Peltier devices to control the temperature, and in oxygenated saline. Slices, once cut, must be held in oxygenated saline. Use Nomarski optics on an upright (as opposed to inverted) microscope to visualize the preparation. Use a gentle stream of saline to remove debris from the surface of the thin slice. Enzymatic treatment is not used, and the method also has the advantage that some of the synaptic connections between cells are preserved.

4.2 Membrane vesicles

A number of laboratories use membrane vesicles. Some have used freeze/thawing of microsomes (19). The most widely used vesicle preparation appears to be that for skeletal muscle. A protocol for amphibian muscle is given in ref. 20, while a variant used for mammalian muscle is given in ref. 21.

4.3 Cells in primary culture

Sometimes it is desirable to work on cells in primary culture. Such work has been carried out on neurons in a number of laboratories. Neurons may be grown on a confluent feeder layer of astrocytes since they do not survive well without a feeder layer of this kind. The following method, developed in Leicester by I. D. Forsythe (see also refs 22 and 23), is convenient (*Protocol 1*).

Protocol 1. Culturing neurons for patch clamping

Preparation of astroglial feeder layer:

1. Dissect the hippocampus from newborn (1–2 day) rats that have been humanely killed (e.g. by decapitation). Dissect off the meninges, which are the main source of fibroblasts.

Protocol 1. *Continued*

2. Wash with Ca^{2+}, Mg^{2+}-free medium. Dissociate with trypsin 0.25% in Ca^{2+}, Mg^{2+}-free medium. Triturate in MEM (minimum essential medium with Earle's salts) supplemented with fetal calf serum.

3. Plate out in culture dishes or 13 mm glass coverslips coated with collagen and polylysine.

4. Feed every 2–3 days with MEM supplemented with fetal calf serum, until the cells become confluent.

5. Add an antimitotic drug to stop further division, and transfer to MEM supplemented with horse serum.

6. Half-feed weekly until neurons are plated.

Plating of neurons

1. Dissect the desired area from the brains of newborn rats that have been humanely killed. It is possible to dissect neurons from known nuclei if the brain is first sliced into 400 μm slices using a Vibratome.

2. Dissociate with trypsin or papain. Triturate in MEM supplemented with horse serum.

3. Plate the neurons on to the glial cells at a density of 50–60 000 cells per 16 mm well.

4. Make a complete change of medium after 24 h, with an antimitotic drug included.

5. Half-feed cells weekly thereafter.

4.4 Cell lines

A number of cell lines may be grown up for use with single channel patch clamping. Various neuroblastoma cell lines have been used in studies of neuronal channels, generally after differentiation, for example, with prostaglandin E_1 and theophylline (24). As with all cells kept in cell culture medium, it is essential to wash the medium off thoroughly before attempting to make seals for patch clamp.

5. Preparing patch pipettes

5.1 Electrode glass

Patch electrodes are pulled from glass capillary tubing 1–2 mm in outside diameter. Standard (thick-walled) 1.5 mm tubing has a wall thickness of about 0.3 mm, and is well-suited to making electrodes for single channel recording; thin-walled tubing with a thickness around 0.16 mm gives lower resistance electrodes which are often preferable for whole cell studies (see Chapter 2).

Most people now use tubing with an internal glass filament, which greatly aids electrode filling. Glass is normally supplied in pieces 10–15 cm long, and needs to be cut to about half this length before pulling. Lightly flaming the cut ends reduces damage to the electrode holder. Several different types of glass may be used for patch electrodes:

- Soft glass (soda glass). This pulls at low temperatures, and makes large pipettes of low series resistance easily. The high conductivity of soft glass gives relatively noisy recordings, however, so that it is usually used for whole cell recording. It can be bought in the form of haematocrit tubes.

- Hard glass (borosilicate or Pyrex). This pulls at higher temperatures, but produces quieter pipettes than soft glass. Thick-walled capillaries are suitable for single channel, and thin-walled for whole cell recording. This is the most commonly used type, and includes normal microelectrode glass.

- Aluminosilicate glass. This needs higher temperatures for pulling than borosilicate, but makes lower noise electrodes.

- Kovar glass. This also needs high temperatures, but produces low noise pipettes that seal well (25). It is not suitable for whole cell electrodes.

Further information of the properties of electrode glass in relation to patch recording may be found in ref. 26.

5.2 Pulling pipettes

Patch pipettes are made using a two-stage pulling process to produce a shorter, stubbier electrode than can be made with a single pull. The first pull thins the capillary over 7–10 mm and the second separates it to form two electrodes. The procedure is described for a vertical puller in *Protocol 2* using gravity to provide the pull. Essentially the same method can be used with a horizontal puller, with the difference that the pull will be provided by an electromagnet.

Protocol 2. Pulling the patch clamp pipettes

1. Pull using high heat, placing a stop so that the moving clamp of the puller is only able to drop by 7 mm. Some pullers have a swing-in stop, but the stop can be as simple as a block placed beneath the moving clamp.

2. Switch the heat off after this first pull (if this is not done automatically). Allow the glass to harden.

3. Slide the glass up to recentre the thinned section in the coil.

4. Remove the stop so that the moving clamp can drop freely, and pull on a lower heat. The heat on the second pull sets the final tip size, more heat giving a smaller tip. A tip diameter of 1 µm or less is desirable.

Protocol 2. *Continued*

Notes

(i) With a vertical puller any magnetic pull should be set to zero. Weight can be added to the moving carriage if necessary.

(ii) Some pullers allow the heater current for the first and second pull to be set separately. This is very convenient. It is possible to modify certain other pullers so that the control normally used to control the pull strength is used instead for the second heat.

(iii) Using a digital voltmeter to monitor the voltage across the heating coil can help to pull consistent electrodes.

5.2.1 Checking electrode size by 'bubble number'

A simple means of checking pipette size after pulling (or after fire-polishing) is to measure the pressure needed to expel air from the pipette into methanol. Connect the back of the electrode to a polythene tube connected to a 10 ml syringe with the plunger withdrawn to the 10 ml mark. Dip the pipette tip into clean methanol in a small vial and press the plunger until a stream of bubbles is driven from the tip. The volume at which this occurs is the bubble number. Illumination from the side and a dark background helps to see the bubbles.

Suitable bubble numbers after pulling are 4–6 ml for single channel electrodes. It is not usually necessary to check every electrode, but the method can be very useful when setting up a puller, or when trying to fire-polish to a particular size. A further check on electrode size is obtained when the electrode resistance is measured before making a seal (see *Figure 6*). With normal pipette and bath solutions, electrodes for whole cell recording have resistances of 2–10 MΩ, while single channel electrodes are typically between 8–40 MΩ.

5.2.2 Electrode pullers

Fairly simple vertical pullers, such as the Kopf 700D and Narashige PP-83, are well-suited for making patch electrodes. More sophisticated pullers can also be used, with some programmable models having facilities for remembering many different pulling protocols.

5.3 Coating

Electrodes for single channel recording, and often also for whole cell (see Chapter 2), are coated with an insulator to reduce electrode capacitance and hence their noise level. Sylgard resin is most often used, though some workers use dental wax. Sylgard is a two-component resin that may be premixed and stored in a freezer in small vials or syringes. It cures rapidly on heating. It is convenient to hold the electrode in a groove in a Perspex or metal block under a spring clip, allowing it to be rotated as Sylgard is applied using a fine syringe needle or wire, and to be withdrawn into a heated wire coil for curing (*Figure 4*).

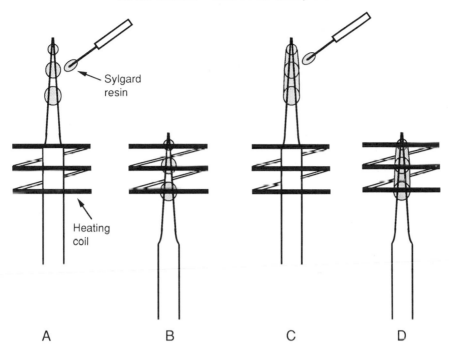

Figure 4. Coating electrodes with Sylgard. A. Apply several drops of Sylgard resin along the length of the electrode, including one near the tip. B. Withdraw the electrode into the heating coil for a few seconds to cure the resin. C. Add resin to fill in between the first blobs. Sylgard may also be added nearer the tip if you wish to coat very near the tip for minimum noise. D. Cure the finished coating.

5.4 Fire-polishing

It is usual to fire-polish electrode tips using a microforge to improve their sealing properties, though sometimes sufficient polishing occurs in the pulling process. Most microforges are homemade, consisting of a platinum wire (~ 50–$100\,\mu m$ diameter) heated by a transformer connected to a Variac to give a final voltage of 0–6 V. The wire is mounted on the stage of a compound microscope (about $\times 400$ magnification) and to prevent evaporation of metal on to the electrode tip, a coating of electrode glass is applied to the wire by melting an electrode on to it. The tip of the electrode to be polished is brought close to the coated wire using a simple micromanipulator. The heat is also localized by passing a stream of air over the wire and electrode, so that the final colour of the heated wire is a dull red. The arrangement is shown in *Figure 5*. The degree of polishing depends on the desired final electrode size, but typically a few seconds is sufficient. It is often possible to see a slight change in the electrode tip as polishing occurs. Electrodes with bubble numbers of 2–4 ml after polishing are usually suitable for single channel recording.

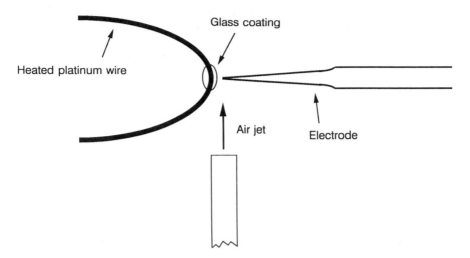

Figure 5. Arrangement of microforge for fire-polishing. Appropriate pressure for the air jet may be obtained by bubbling air from a pump or air line into a few centimetres of water, and taking the supply for the jet from a side branch of the tube used to do this.

5.5 Filling

Clean filling solutions by filtering through a 0.2 μm Millipore filter before use. Electrodes made from filament glass (see Chapter 1) may simply be back-filled by inserting a fine (30-gauge) syringe needle or a Pasteur pipette (pulled into a fine tube) as far as possible into the electrode and then delivering solution to fill the shank about half-way. Non-filament glass electrodes should be filled as follows:

- connecting the back of the electrode to a fine polythene tube connected to a 5–10 ml syringe.
- dip the electrode tip into a small beaker of electrode solution
- use the syringe to apply suction for 30–60 s, this will fill the electrode tip
- back-fill the electrode shank as above

Any remaining bubbles (which usually occur where the electrode begins to taper) can be seen most easily by illuminating the electrode from the side and viewing against a black background. They can often be removed by flicking the electrode shaft with a fingernail.

5.6 Junction potentials

When the solutions in the patch pipette and bath are of different ionic composition a liquid junction potential will develop at the pipette tip because of differences in the mobilities of different ions. This junction potential will be present when the pipette current is zeroed with the patch clamp amplifier before making the seal, but will disappear when a seal is made. Between physiological KCl and NaCl solutions, the junction potential will be quite small (<4 mV), but when lower mobility anions, such as aspartate or gluconate, are used to substitute for some of the Cl^- in pipette solutions for whole cell recording, junction potentials can be as big as -10 mV (pipette relative to bath).

5.6.1 Measuring the junction potential

Junction potentials can be measured by using a 3 M KCl electrode as a reference electrode, since this develops negligible junction potential itself. The most stable and fastest response is obtained with a flowing KCl electrode (see, for example, ref. 27), but a piece of electrode glass filled with 3 M KCl agar can also be used. The procedure is to dip both reference electrode and pipette into a small dish filled with the pipette solution, and to zero or record the (zero current) voltage between them using the current clamp mode of a patch clamp amplifier. Under this condition there will be no junction potential. The electrodes are then moved to a dish containing the bath solution and the resulting junction potential measured as the voltage difference between this and the previous solution, expressed pipette relative to bath.

5.6.2 Correcting for the junction potential

Once a seal has been made, the true potential in the patch pipette will be given by its recorded value plus the junction potential measured above. Thus if the junction potential was -10 mV (pipette relative to bath) the true membrane potential of a cell under whole cell clamp will be -60 mV when the holding potential is set to -50 mV.

6. Making the seal

6.1 Earthing the bath

The bath solution surrounding the preparation needs to be connected to the earth terminal on the headstage of the patch clamp amplifier. The simplest bath electrode is a silver/silver chloride (Ag/AgCl) pellet or chlorided silver wire placed directly in the bath. This risks some contamination of the preparation with silver ions, however, and will also give changes in junction potential if bath $[Cl^-]$ is changed. Therefore many people prefer to put the Ag/AgCl electrode in a separate pot of saline connected to the bath by a saline–agar bridge.

6.2 Prerequisites for making seals

Two requirements for seal formation are:

- Clean solutions. All solutions should be filtered with a 0.2 μm Millipore filter.
- A system for applying suction to the electrode by way of the tube provided on the electrode holder. The simplest and most widespread method is to connect a length of polythene or silicone tubing ending in a 1 ml syringe barrel

or disposable pipette tip, to which suction is applied by mouth. A two-way syringe tap, mounted in line so that it can be closed to maintain suction, is also useful. Alternatively, a syringe or a vacuum pump connected to a water manometer can be used to apply suction. **Caution:** of course, this latter system should be used if the pipette solution is toxic.

6.3 Seal formation

Details of the method used to form pipette–cell seals are given in *Protocol 3*.

Protocol 3. Sealing the pipette to the cell membrane

1. Mount the pipette in the pipette holder, screwing the end up tightly so that an airtight seal is formed and the pipette does not move when suction is applied.

2. Set the patch clamp amplifier to search mode, and the gain to about 50 mV/pA (this can be adjusted as needed to obtain a convenient display). Apply a small command voltage (typically 0.5 or 1 mV) to the amplifier to generate a current through the pipette (*Figure 6A*). This is used to monitor the pipette resistance and thus seal formation. Pipette resistance is given by command voltage divided by pipette current.

3. Apply slight positive pressure to the pipette and lower the tip through the air–solution interface. Positive pressure helps keep the tip clean; some people keep a steady positive pressure on until the pipette touches the cell.

4. Position the pipette tip close to the cell, and advance it very slowly until it touches. An increase in resistance of about 10–80%, seen as a corresponding decrease in pipette current (*Figure 6B*), occurs as the touch is made.

5. Apply gentle suction to the pipette, watching the pipette current. Depending on the preparation, the seal may form slowly and progressively or abruptly. Switching the amplifier to voltage clamp and applying a holding voltage of up to 60 mV may help in the later stages of sealing. The formation of a seal is indicated by the abolition of any visible current step in response to the test voltage, and by a large reduction in noise (*Figure 6C*). The final seal resistance may be measured by increasing the size of the command voltage pulse to 10–50 mV, and increasing the gain on the amplifier until the resulting step in the pipette current can be measured. This measurement can be hard for very good seals, for example a 20 mV pulse will only give 0.2 pA of current through a 100 GΩ seal. On the other hand, the exact resistance does not matter for seals this good.

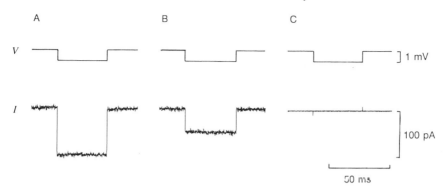

Figure 6. Making a seal. The example shows the current (lower trace) through a 10 MΩ electrode in response to a 1 mV test pulse. A. The electrode tip is in the bath, but not touching the cell. The 1 mV pulse gives a 100 pA current through the electrode. B. As the electrode touches the cell, the current becomes smaller. C. Suction is applied to make the seal. Formation of a gigaohm seal is indicated by the disappearance of a visible current pulse and the reduction in noise.

6.4 When seals won't form

Usually blame is placed either on the preparation or the electrodes. Here are some things to try:

- Check that the suction system will hold steady suction. If it won't, the pipette holder may not be done up tightly enough, or the O-ring may be damaged.
- Check that the anti-vibration table is inflated.
- Replace the glass coating on the wire of the fire-polisher.
- Make a few new electrodes, and try these.
- Try a new dish of cells.
- Re-filter solutions. Check that there is no grease or cell debris on the surface of the bath solution.
- If someone nearby has cells (or electrodes) which are sealing well, borrow some and try them on your set-up.
- It is a common experience that seals just won't form on some days, with some preparations. Sometimes the only solution is to relax and try again the next day. Careful observation of someone who has a working set-up can be the most helpful thing of all.

7. The variants of patch clamp recording

There are two basic variants of patch clamp recording—methods that attempt to record from single channels (or from a handful of channels) in a small patch of membrane, and methods that allow the experimenter to record from the whole cell. The methods are generally known as *single channel recording* and *whole cell recording* respectively. The method of single channel recording itself has a number of variants, for recording may either be carried out with the patch of membrane still remaining part of the cell (*cell-attached recording*) or with the patch torn off the cell as a cell-free, *excised membrane patch*. Excised patches may themselves be made in an *inside-out* (cytoplasmic face outwards) or an *outside-out* configuration (extracellular face outwards). In using cell-attached patches, especial attention needs to be given to the avoidance of mechanical vibration, since movement between the recording pipette and the cell may itself excise the membrane patch. Excised patches are often used to study the effects of agents applied either from the intracellular or the extracellular medium on channel function.

Because currents are small with single channel recording, use *thick-walled* glass for electrode manufacture (see Section 5.2) and coat with Sylgard (see Section 5.3) to reduce current noise.

Conversely, because currents in whole cell recording are often large, so that minimizing current noise is less important than the series resistance errors that may arise from the resistance of the electrode, use *thin-walled* glass for such recording.

In all forms of patch clamp recording it is necessary to keep the convention that inward membrane currents are ascribed a negative sign and that membrane potentials are measured inside relative to outside. Patch clamp amplifiers may have their own convention—for example the List EPC-7 gives current flowing out of the tip of the pipette a positive sign. In all cases, changes of membrane potential are made by changing the potential of the interior of the patch pipette.

8. Single channel recording

8.1 Cell-attached recording

Protocol 4: Cell-attached recording

1. Make a seal as described in Section 6.3, using an appropriate extracellular solution in the patch pipette. If you are using the auto-zero function of the patch clamp amplifier to set the current and voltage zeroes during the making of a seal, switch to voltage clamp after the seal is formed. For example, turn the control from 'search' to 'voltage clamp' for the List EPC-7.

2. If you are recording channels that are activated by a change in membrane potential (voltage-gated channels), you will need to adjust the capacity compensation controls on the patch clamp amplifier. Use a small voltage pulse to make these adjustments (see for example *Figure 7A* and *B*). First adjust the control with the shortest time constant, and subsequently the longer ones. If you are using a computer, digital methods may be used subsequently to subtract the remains of the capacity currents associated with the charging of the pipette capacity, provided these currents have not saturated the input of the A/D converter. Some analog compensation is always necessary to prevent such saturation.

3. Compensate for, where possible, the leakage currents flowing primarily through the resistance of the seal. Some amplifiers (for example the Axopatch) possess a facility for leakage subtraction. In other cases, compensation may be done by subtracting an appropriate fraction of the voltage command from the pipette current. The size of the fraction will be decided by the size of the seal resistance and the gain that is being used to measure current. An instrumentation amplifier such as the Analog Devices AD524 may be used for such subtraction.

4. Turn the gain of the patch clamp amplifier to a suitable level to measure currents as small as 1 pA. If you are using a computer and interface, an overall gain of 1 V/pA allows the use of an appropriate fraction of the input range of the A/D converter. The final choice of gain will be decided by the size of the currents.

5. Remember, when using cell-attached recording, that the membrane potential of the cell will add to the voltage imposed within the patch pipette. Whether the membrane potential remains constant through the experiment can be assessed from constancy of the amplitude of unitary currents. An estimate of the membrane potential may be made *at the end of the experiment* by breaking into the cell (by applying negative pressure to the inside of the patch pipette) and measuring the voltage that needs to be applied to bring the pipette current to zero. If you are unable to measure the resting potential in this way it will be necessary to report membrane potentials as changes from the resting potential.

8.2 Excised patches: inside-out configuration

Inside-out patches are readily made from most cell types.

Protocol 5. Formation of inside-out excised patches

1. Fill the recording pipette with a suitable extracellular solution and seal on to the cell in the usual way.

2. Now pull the recording pipette away from the cell. In most instances the seal between the pipette and the membrane will be sufficiently good that a patch will be pulled off[a].

---

[a]Sometimes patches become completely silent when excised, and the noise in the current record becomes very low. Such silence almost always indicates that the patch has sealed over as a small vesicle in the tip of the pipette, and the reduction in noise is due to the very high resistance now sealing the pipette tip. The chance of this happening is much reduced if patches are excised into a Ca^{2+} free solution.

Moving the tip of the pipette briefly through the interface between the bath solution and air may reopen a vesicle. Some authorities argue that excised inside-out patches should always be taken through the air–solution interface to ensure that they are open and that channel events are recorded at full height. Unhappily, the patch will often fall off the pipette when you do this.

Protocol 5. *Continued*

3. Bring the patch into contact with an appropriate intracellular solution. Isotonic KCl, buffered to an appropriate pH, and with zero Ca^{2+} (with EGTA) is often adequate as an intracellular solution. More complex solutions may be used. Cl^- may be replaced by an organic anion such as gluconate or aspartate, though it will then be important to correct appropriately for the large junction potentials that will arise between the pipette solution and the bath solution (see Section 5.6). GTP and ATP may be added at appropriate concentrations if required.

8.3 Excised patches: outside-out configuration

It is often more difficult to make outside-out patches (*Protocol 6*). Some cell types (adult skeletal muscle fibres, for example) do not lend themselves to this approach, perhaps because of the arrangement of the sub-membrane cytoskeleton.

Protocol 6. Formation of outside-out excised patches

1. To make an outside-out patch, seal on to a cell in the normal way, using a pipette made with thick-walled glass, but filled with a suitable intracellular solution. The tip diameter of the pipette needs to be large enough for it to be possible to break into the cell.

2. Now apply brief suction to break into the cell, using the same method as used for whole cell recording (see Section 9 and *Figure 7*). When the patch of membrane in the pipette tip is disrupted, the record will become noisier, the capacity transients will become larger as the cell capacity is added, and the zero current level will alter because of the resting potential of the cell. You may wish to set a holding potential before switching from 'search' to 'voltage clamp' modes.

3. Pull the electrode away from the cell. Doing this will pull out a 'neck' of membrane, which will then break from the cell. The pulled-out membrane must reseal to form an outside-out patch, and this resealing can be assessed from the electrical resistance in the pipette tip. It is necessary to excise patches into a suitable extracellular medium. The presence of Ca^{2+} in this medium will aid resealing.

Outside-out patches are larger than inside-out ones. The electrical capacity, and hence the current noise, is greater.

As an alternative strategy, the membrane patch first formed can be permeabilized using nystatin (see Section 9.2 and ref. 28). Such a procedure has been shown to retain intracellular components within what is now a permeabilized vesicle. It has the advantage of retaining the second messenger

A

On cell patch

B

Compensated for electrode capacitance

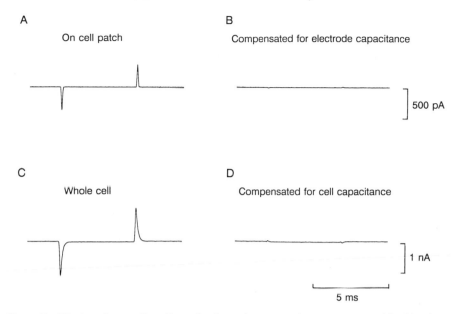

500 pA

C

Whole cell

D

Compensated for cell capacitance

1 nA

5 ms

Figure 7. Whole cell recording. Records show the current in response to a 10 mV voltage clamp step. A. The electrode is sealed on to the cell and the transient capacity currents occur at the beginning and end of the step. B. The electrode capacitance is compensated for using the controls on the patch clamp amplifier. C. A pulse of suction is used to break into the cell. Large transient capacity currents are seen due to the cell capacitance. Note that the gain of the recording has been halved compared to A and B. D. The patch amplifier controls are used to compensate for the cell capacitance.

biochemistry of the cell and is ideal for the study of modulation of channels when unitary currents are measured.

9. Whole cell recording

9.1 Conventional method

Currents recorded during the procedure are illustrated in *Figure 7*, and described in *Protocol 7* (see also Chapter 2, Section 4).

Protocol 7. Whole cell voltage clamping

1. Use a thin-walled pipette filled with an intracellular solution to seal on to the cell.
2. Use the fast capacity compensation of the patch clamp amplifier to compensate for the capacity of the pipette (*Figure 7B*).
3. Set a suitable holding potential so that the cell will not be clamped to zero when the amplifier is switched from 'search' to 'voltage clamp'.

Protocol 7. *Continued*

4. Apply a brief pulse of suction to break into the cell. As an alternative a large voltage pulse or voltage oscillation may be used—the Axopatch amplifier has a 'zap' button that will do this[a]. When the membrane patch is disrupted, a large capacity transient will appear (*Figure 7C*), the current record will become noisier, partly because of the increased capacity and partly because of noise associated with the opening and closing of channels, and the zero current level will change. If you remain in the 'search' mode, the resting potential will be recorded on the voltage meter of the amplifier. Now switch to voltage clamp (having first set a suitable holding potential).

5. Adjust the capacity compensation for the cell capacity[b], and compensate for the series resistance (*Figure 7D*).

[a] After a connection has been formed between the inside of the recording pipette and the interior of the cell, the contents of the cell begin to wash out. The exchange is rapid for the more diffusable intracellular cations and anions, but may occur more slowly for larger molecules and intracellular components. During this washout, the junction potential between pipette contents and cell contents will change and this change may affect recordings of voltage-gated ion channels (see ref. 29). For studies of channel modulations through G-proteins and intracellular second messengers you will need to supply suitable substrates from the pipette solution.

[b] The size of the cell may be estimated electrically from the transient associated with charging its capacity. The transient will relax exponentially in a spherical cell. Its area (in coulombs) will be given by $C_m \cdot V$, where C_m is the membrane capacity (farads) and V is the size of the applied voltage step (volts). The time constant of the exponential relaxation is given by:

$$\tau = R_s \cdot C_m \tag{2}$$

Where R_s is the series resistance, which will usually be greater than (about twice) the pipette resistance as measured in the bath before a seal is made. Most patch clamp amplifiers have a mechanism for compensating for series resistance, adjusting the voltage pulse to ensure that the desired voltage is applied across the cell membrane resistance, R_m, rather than across $(R_s + R_m)$. The presence of the series resistance will limit the time resolution of membrane currents, with an approximate bandwidth,

$$f_c = 1/(2\pi\tau) \tag{3}$$

Whole cell recording from neurons may also be affected by poor voltage control in processes such as neurites.

9.2 Alternative strategy: use of nystatin

Recently Horn and Marty (30) have introduced a method of whole cell recording which uses the ionophore (and antibiotic) nystatin to permeabilize the patch of membrane isolated with the pipette tip. This ionophore is permeable to small cations such as Na$^+$ and K$^+$. Exchange of these, therefore, occurs between the pipette and cell, but other cellular contents are conserved. The choice of this method as opposed to the more classical methods of whole cell recording depends on the nature of the experiment to be performed. Effective recording is slightly more difficult to achieve

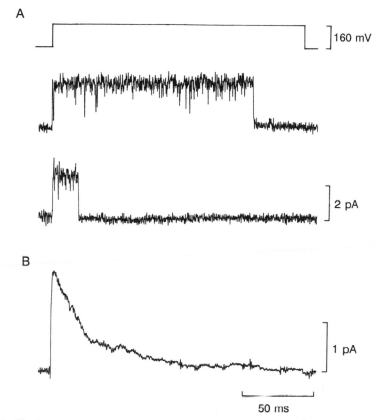

Figure 8. Single channel recordings of A-currents. A. Two examples of single channel recordings. Opening of the A-current channel occurred in response to a voltage step (shown above) from the holding potential of -100 mV to $+60$ mV. The pipette contained 3 mM K^+, while the solution bathing the cytoplasmic face of the patch contained 140 mM K^+. Filtered at 2 kHz (8-pole Bessel) and sampled at 10 kHz. B. Ensemble average of 100 records like those of A, showing the transient nature of the averaged current.

than with the conventional method. First, the nystatin technique is a bit more fiddly. Secondly, the resistance of the nystatin-permeabilized patch will always be rather greater ($2-3\times$) than that achievable by breaking the patch. Because the monitoring of capacity is important in assessing the connection between the pipette and the cell, coat the electrodes to reduce electrode capacity. The following method is used (*Protocol 8*):

Protocol 8. Membrane permeabilization with nystatin

1. Fill the tip of a thin-walled pipette with nystatin-free intracellular solution by dipping the pipette tip into the solution for a few seconds. Backfill the pipette with intracellular solution containing nystatin at

81

Protocol 8. *Continued*

20–100 µg/ml. The nystatin should be added from a stock solution made up in methanol. It will be necessary to use an ultrasonicator to keep the nystatin in solution as it is diluted into the pipette solution.

2. Seal on to a cell in the usual way, and compensate for the electrode capacity. As nystatin gradually permeabilizes the membrane patch, the capacity transient will increase in area and the series resistance will fall. Monitor these quantities as the connection is formed. It may take a few minutes for permeabilization to occur. Nystatin is cation-selective, and though it is also weakly permeable to Cl^- most intracellular anions will not permeate. As a result a junction potential, associated with the Donnan equilibrium that will arise because of the presence of impermeant anions within the cell, will develop between pipette and cell contents. An account of this potential (which will be several mV) is given in ref. 30.

3. After permeabilization has reached its equilibrium, switch to voltage clamp with an appropriate holding potential, compensate for capacity transients and series resistance in the usual way, and proceed with the experiment.

10. An example of single channel recording

Figure 8 shows potassium currents recorded using an excised inside-out membrane patch. The channel isolated is one that carries a transient outward current often called A-current. This current helps to control the interval between action potentials in neurons. The recording is from a neuron from the rat locus coeruleus grown in primary culture using the method outlined in Section 4.3.

Acknowledgements

We thank Drs N. W. Davies and I. D. Forsythe for their comments on this chapter, and for providing material for Figs 6, 7, and 8.

References

1. Neher, E. and Sakmann, B. (1976). *Nature*, **260**, 799
2. Hamill, O. P., Marty, A., Neher, E., Sakmann, B., and Sigworth, F. J. (1981). *Pflügers Arch.*, **391**, 85.
3. Ogden, D. C. (1987). In *Microelectrode Techniques. The Plymouth Workshop Handbook*. (ed. Standen, N. B., Gray, P. T. A., and Whitaker, M. J.), pp. 199–227. Company of Biologists, Cambridge.
4. Ogden, D. C. and Stanfield, P. R. (1987). In *Microelectrode Techniques. The Plymouth Workshop Handbook*. (ed. Standen, N. B., Gray, P. T. A., and Whitaker, M. J.), pp. 63–81. Company of Biologists, Cambridge.
5. Sakmann, B. and Neher, E. (ed.) (1983). *Single Channel Recording*. Plenum Press, New York.

6. Sigworth, F. J. (1983) In *Single Channel Recording*. (ed. Sakmann, B. and Neher, E.), pp. 3-35. Plenum Press, New York.
7. Lamb, T. D. (1985). *J. Neuroscience Meth.*, **15**, 1.
8. Colquhoun, D. (1987). In *Microelectrode Techniques. The Plymouth Workshop Handbook*. (ed. Standen, N. B., Gray, P. T. A., and Whitaker, M. J.), pp. 83-104. Company of Biologists, Cambridge.
9. Colquhoun, D. and Sigworth, F. J. (1983). In *Single Channel Recording*. (ed. Sakmann, B. and Neher, E.), pp. 191-263. Plenum Press, New York.
10. Freshney, R. I. (1983). *Culture of animal cells. A manual of basic technique*. Alan R. Liss, Inc., New York.
11. Bottenstein, J. E. and Sato, G. (1985). *Cell culture in the neurosciences*. Plenum, New York.
12. Shahar, A., de Vellis, J., Vernadakis, A., and Haber, B. (1989). *A dissection and tissue culture manual of the nervous system*. Alan R. Liss, Inc., New York.
13. Kay, A. R. and Wong, R. K. S. (1986). *J. Neuroscience Meth.*, **16**, 227.
14. Powell, T., Terrar, D. A., and Twist, V. W. (1980). *J. Physiol.*, **302**, 131.
15. Langton, P. D., Burke, E. P., and Sanders, K. M. (1989). *Am. J. Physiol.*, **257**, C451.
16. Warshaw, D. M., Szarek, J. L., Hubbard, M. S., and Evans, J. N. (1986). *Circ. Res.*, **58**, 399.
17. Trube, G. (1983). In *Single Channel Recording*. (ed. Sakmann, B. and Neher, E.), pp. 69-76. Plenum Press, New York.
18. Edwards, F. A., Konnerth, A., Sakmann, B., and Takahashi, T. (1989). *Pflügers Arch.*, **414**, 600.
19. Tank, D. W., Miller, C., and Webb, W. W. (1982). *Proc. Natl. Acad. Sci. USA.*, **79**, 7749.
20. Standen, N. B., Stanfield, P. R., Ward, T. A., and Wilson, S. W. (1984). *Proc. Roy. Soc. Lond. B.*, **221**, 455.
21. Burton, F. L., Dörstelmann, U., and Hutter, O. F. (1987). *J. Physiol.*, **392**, 12P.
22. Forsythe, I. D. and Westbrook, G. L. (1988). *J. Physiol.*, **396**, 515.
23. Masuko, S., Nakajima, Y., Nakajima, S., and Yamaguchi, K. (1986). *J. Neurosci.*, **6**, 3229.
24. Higashida, H. and Brown, D. A. (1986). *Nature*, **323**, 333.
25. Rae, J. L. and Levis, R. A. (1984). *Biophys. J.*, **45**, 144.
26. Corey, D. P. and Stevens, C. F. (1983). In *Single Channel Recording*. (ed. Sakmann, B. and Neher, E.), pp. 53-68. Plenum Press, New York.
27. Spitzer, K. W. and Walker, J. L. (1979). *Pflügers Arch.*, **382**, 281.
28. Levitan, E. S. and Kramer, R. H. (1990). *Nature*, **348**, 545.
29. Marty, A. and Neher, E. (1983). In *Single Channel Recording*. (ed. Sakmann, B. and Neher, E.), pp. 107-122. Plenum Press, New York.
30. Horn, R. and Marty, A. (1988). *J. Gen. Physiol.*, **92**, 145.

4

Video imaging of neuronal activity

MICHAEL M. HAGLUND and GARY G. BLASDEL

1. Introduction

Recent studies of the central nervous system have combined different techniques in an attempt to understand the relationship between anatomy and physiology. The functional organization of the cortex has become much more familiar through advances in imaging techniques such as 2-deoxyglucose autoradiography (DG), Computerized Tomography (CT), Magnetic Resonance Imaging (MRI), Positron Emission Tomography (PET), and Single Photon Emission Computerized Tomography (SPECT). The idea of direct optical imaging has occurred to many workers and was first tried more than 40 years ago (1), but has resurfaced recently on account of the many advantages that it has over other techniques.

Most of our current understanding of functional organization in the neocortex was painstakingly acquired through thousands of microelectrode recordings at various sites. These have provided intriguing insights, especially in conjunction with anatomical approaches (2,3). However, the ability of microelectrode recordings to resolve two-dimensional organization of activity over large areas is restricted by the time needed to obtain recordings of adequate quality at each site.

Metabolic labelling with radioactive DG, a hybrid of physiological and anatomical techniques (4), provides one means of visualizing cortical activity in two dimensions. Recently, PET and SPECT scanning have been exploited to reveal many details of functional anatomy, especially when used in combination with high resolution MRI (5,6). These techniques have the advantage over DG that they may be used interactively and repeatedly in the same tissue. Because they are relatively non-invasive, they may also be used in humans. Their only real limitation is poor resolution, spatial and temporal. Newer generations of scanners will undoubtedly improve on existing values, but only to a point because the fundamental laws of nature preclude spatial resolution better than 2 mm.

Because the temporal and spatial resolution of light is limited only by its speed and wavelength, optical recording of neuronal activity offers significant advantages in this regard. By applying voltage-sensitive dyes to the surface of the brain, or by using the appropriate wavelength of light to illuminate the brain surface, active regions of cortex change their absorbance and become darker. This darkening of the cortex can be visualized with photodiode arrays or by video cameras. Optical imaging of neuronal activity can provide spatial resolution as small as 2 µm with temporal resolution of less than 1 ms.

The potential uses of optical recording have been advanced during the last 20 years by the work of a small group of investigators that include (but is not limited to) Cohen, Waggoner, Davila, Salzberg, Grinvald, Ross, and Orbach. These investigators have made vital contributions to the technique by screening and developing optical probes of membrane potential (voltage-sensitive dyes) and by pioneering different methods for their use. Blasdel and Salama (7) expanded on these technologies by using a television camera to obtain greater spatial resolution (120 × 100) than previously had been possible with photodiode arrays (12 × 12). These investigators have used optical imaging of neuronal activity in preparations ranging from Aplysia ganglion cells, invertebrate nerve terminals, brain slices, and monkey visual cortex. There are a vast number of questions addressable with the optical imaging technique.

The breadth of recent developments in optical imaging, with and without voltage-sensitive dyes, has been reported recently (8–10). This chapter concentrates on the specific method used in video imaging of neuronal activity in this laboratory. The chapter will initially discuss the sequence of animal (monkey) preparation including detailed protocols on animal set-up, surgery/chamber implantation, long-term maintenance/monitoring, and post-operative recovery. Section 3 describes the optical imaging equipment we currently use, while Section 4 focuses on the fundamentals of optical imaging with and without voltage-sensitive dyes (without dyes refers to imaging the natural changes in the brain tissue that occur with neuronal activity, referred to as the intrinsic signal). Section 4 also discusses the source of the voltage-sensitive dye signals, dye application, and sources of noise. Section 5 demonstrates basic protocols for differential video imaging and describes a number of applications focusing on solutions to problems, imaging paradigms, and potential pitfalls.

2. Animal preparation

These experiments depend on careful surgical preparation and thoughtful maintenance of each animal. These are easily the most important parameters in the successful performance of any study. Work in this laboratory has concentrated on Macaque monkeys (*Macaca nemestrina* and *Macaca cynomolgus*) because the striate cortex in this species is better understood than any other area of cortex in any animal (11). It is ideal for optical studies because it is large, flat, and conveniently located on the surface (at the back of the animal's head) where it is readily accessible. We tend to use small monkeys (1.5–3.0 kg) due to ease of handling and the lower cost for experimental drugs. The cost of drugs for each experiment is in the range of US$75–100 and will be lower for smaller animals because most drugs are given per kg body-weight.

The maintenance of any animal in an anaesthetized, paralysed state, where the cells in primary visual cortex remain functional, is itself difficult; a demanding exercise in experimental physiology and pharmacology entailing many more variables than can reasonably be covered here. This section is intended, therefore, to demonstrate our strategies through a number of protocols detailing animal set-up, surgery and chamber implantation, long-term maintenance and monitoring of the animal during a 24-hour procedure, and recovery.

You will need the following equipment for *Procotol 1*:

- dexamethasone (Decadron)
- xylazine (Rompum)
- chloramphenicol (1 g)
- ampicillin (1 g)
- d-tubocurarine
- ketamine
- potassium chloride (40 mM)
- thiopentone (5 mg/ml)
- Isolyte or balanced salt solution for i.v. fluids
- vecuronium (Norcuron) (0.25–0.33 mg/ml)
- syringes—3, 5, 10, 20 ml tuberculin (1 ml)—for mixing and administering drugs
- 21 or 22 gauge intravenous catheter
- adhesive tape for securing i.v. and urinary catheters, endotracheal tube
- shaver
- paediatric feeding tube (for urinary catheter)
- laryngoscope
- cuffed endotracheal tube (3-0 or 4-0)
- stereotaxic apparatus
- small animal ventilator
- air table
- refrigerator (for solutions)
- thermistor (either oesophageal or rectal)
- temperature control unit and power supply for heated blanket
- Cole Parmer, Masterflex pumps
- Razel A-series continuous injection pumps

Protocol 1. Animal set-up

1. On the evening before each experiment, fast the animal and give a prophylactic intramuscular (i.m.) injection of dexamethasone (0.15 mg/kg) to decrease brain swelling.

2. Repeat similar doses of dexamethasone every six hours until completion of the experiment.

3. On the morning of the experiment, anaesthetize the animal with ketamine (10–20 mg/kg) and xylazine (0.25 mg/kg) i.m.

4. After shaving an arm or leg, place an i.v. line (20–22 gauge) for fluid administration (Isolyte with 40 mM potassium chloride and antibiotics— ampicillin and chloramphenicol (both 250 mg/kg in 500 ml Isolyte)); typical maintenance i.v. fluid rates are 1–2 ml/kg/h. By placing the i.v. line in the first 5–10 minutes after induction, further doses of ketamine are not necessary because small doses of i.v. thiopentone (1–2 ml) can be given (thiopentone solution is normally in a concentration of 5 mg/ml). We have found that large doses of ketamine during the set-up period limit the quality of the microelectrode and optical recordings. We opt for the short acting i.v. thiopentone instead in order to maintain analgesia while completing the preparation.

5. Catheterize the animal with a urinary catheter consisting of a small paediatric feeding tube. Drain the urine to a urimeter that allows measurement of the urine output. Appropriate fluid administration should result in total input of i.v. fluids balanced with urine output; the urine output should be no less than 0.5 ml/kg/h.

6. Place the animal on a heated waterbed, cover it with a temperature controlled heated blanket, and position in the stereotaxic apparatus.

7. Just prior to intubation give a bolus of i.v. thiopentone to decrease the gag reflex and, therefore, ease access to the vocal cords for intubation.

8. Intubate with a 3-0 or 4-0 cuffed endotracheal tube. Take care to secure the endotracheal tube with tape so it will not be dislodged during the experiment.

9. Insert a temperature probe (YSI 700, Cole Parmer) in the oesophagus to monitor the animal's core temperature. The temperature probe is connected to a controller (Cole Parmer) that drives the heated blanket and waterbed. Temperatures should be kept in the range of 36–37 °C because lower temperatures are related to decreased activity in the cortex and poor quality optical signals.

10. Ventilate the animal with a 2:1 nitrous oxide/oxygen mixture. The respiratory rate should be in the range that combined with the normal tidal volume (10–15 ml/kg) controls the end-tidal CO_2 in the range of

36–40 mmHg (see exceptions to this range below). Normally for small monkeys (1.5–3 kg) we ventilate at 35–40 breaths/minute and tidal volumes of 25–35 ml/breath.

11. Anaesthesia is maintained with i.v. thiopentone (0.5–2.0 mg/kg/h) either as a continuous infusion or intermittent bolus dose of 1–2 ml.

12. At the time of eye refraction (see below), paralyse the animal with d-tubocurarine (5 mg/kg) before starting a constant infusion of vecuronium (0.01–0.02 mg/kg/h). Vecuronium has a very short half-life such that at the end of the experiment the animal can be more easily weaned from the ventilator.

You will need the following equipment for *Protocol 2*:

- sterile surgical instruments (ROBOZ): No. 11, 22 scalpels, Metzenbaum scissors, straight 3-0, 4-0 curettes, trephine, fine dural scissors or angled Potts, fine forceps, dural hook, Penfield No. 4, dental freer, needle holders
- dental drill (with assortment of burrs) and irrigator (normal saline)
- bupivacaine (Marcain) 0.5% containing 1:200 000 adrenaline
- contact lenses (full set gas-permeable 5.0–8.0 mm diameter)
- contact lens cleaner
- cotton swabs, gauze pads
- spot retinoscope
- stainless steel chamber with inserts (cap, video imaging, and microelectrode recording)
- durafilm (Codman)
- shaver
- betadiene brush
- dental cement
- sutures (2-0 and 4-0 silk sutures)
- silicon vacuum grease (for sealing inserts in chamber)
- assortment of O-rings for chambers (No. 16–20)
- gentamicin (Garamycin) ointment
- chloramphenicol ointment
- TV monitor

Protocol 2. Surgery and chamber implantation

1. Shave the back half of the scalp and scrub with a betadiene solution for 1–3 minutes.

Protocol 2. *Continued*

2. Inject the line of incision (linear in the midline along the sagittal suture) which extends from 2 cm behind the eyebrows to 2 cm past the posterior portion of the skull with 0.5% bupivacaine with 1:200 000 adrenaline. The bupivacaine provides long-lasting pain relief while the adrenaline constricts blood vessels in the scalp minimizing bleeding.

3. Reflect the scalp using a No. 22 scalpel and small curettes. Sutures can be used to hold the scalp in a more open position.

4. Use a 25 mm diameter trephine or dental drill with a large cutting burr to bore a hole just behind the lunate sulcus, as close as possible to the midline (*Figure 1A*). The midline needs to be avoided because the sagittal sinus lies in the dura at the midline and severe (sometimes life-threatening) bleeding will result if the sinus is damaged.

5. Using a dental drill with a small burr, make multiple holes through the bone edge. Open the dura in a pinwheel fashion and use a 4-0 silk suture to tack the dura to the bone edge (*Figure 1A*). Place a circular piece of durafilm, to extend just beyond the bone edges, between the dura and the surface of the brain. This will protect the brain when fixing the chamber to the skull.

6. Insert a stainless steel chamber, 25 mm in diameter, with roughened outer edges (for a better grip of the dental cement) and an inner threaded portion (for a holding ring with glass coverslip during imaging, or insert with O-ring to seal to glass on microelectrode carrier) and cement to the cranium with dental cement (*Figure 1B, 1C*). Three additional small screws can be placed in the skull to give the dental cement a better anchor. The dental cement should be very smooth to prevent irritation of the scalp which will rub against it after the procedure.

   This chamber can be used either for microelectrode or for optical recordings. In the case of electrode recordings, a thin O-ring, inserted between the chamber and a thin glass disk, secures the system and prevents leakage of physiological solutions. For the optical recordings, the system is sealed by a threaded stainless steel plug, equipped with an 18 mm diameter window (made from a glass coverslip) and two 18 gauge ports to allow superfusion of the cortical surface with the solutions containing the voltage-sensitive dye.

7. At the time of microelectrode recordings, inject the animal with d-tubocurarine (1–2 mg/kg) i.v. to paralyse the eye movement and fit the gas-permeable lenses. Using a spot retinoscope, refract the eyes so that the animal is focused on a television screen 2 metres away. The TV is used for displaying the stimulus patterns (primarily drifting gratings at different angles at a speed of 0.5–3.0 metres/minute. (See later section on practical aspects of differential imaging for further details.)

8. At the conclusion of the experiment, cover the cortex with a disc of durafilm. A cap that fits the chamber can then be inserted (*Figure 1D*). We use

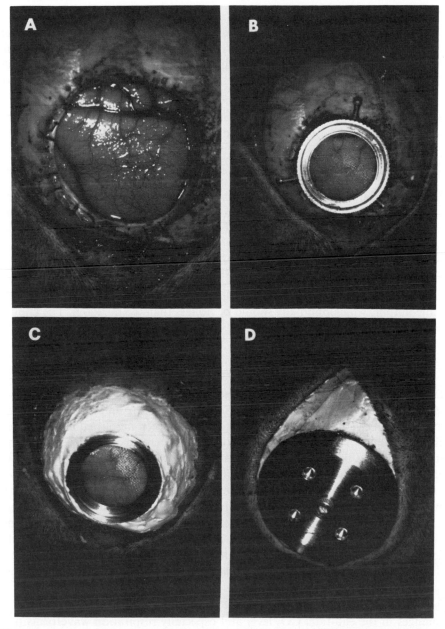

Figure 1. Surgery. Four steps in the surgical procedure. The anterior of the skull is at the top in all figures. (A) After the scalp is removed, holes are made in the skull and the dura sutured to the bone. This prevents the dura from adhering to the cortical surface. (B) Chamber awaiting dental cement. Durafilm has been placed over the cortical surface to protect it from the dental cement. (C) After dental cement is applied. (D) Closure of the chamber showing the cap and small screw used to relieve pressure when inserting cap. The scalp has also been sutured.

Protocol 2. *Continued*

   antibiotic ointments containing gentamicin and chloramphenicol on the inner
   surface of the cap to help prevent infection. The cap should have a small
   central hole to allow fluid to escape otherwise intracranial pressure will
   increase dramatically and may harm the underlying brain tissue. Once the fluid
   has escaped a flat head screw may be placed to seal the chamber (*Figure 1D*).

9. Approximate the reflected scalp using 2-0 silk sutures. The sutures and the
   knots should be well hidden in order to prevent the animal from picking
   at the suture line.

Experiments last 18–24 hours, during which time the animal is repositioned every
2–4 hours to prevent fluid build-up and pressure sores. The fluid balance, heart
rate, blood pressure, end-tidal CO_2, EEG, and level of paralysis all need to be
carefully monitored.

2.1 Monitoring fluid balance

Maintain fluid balance by adjusting the input of i.v. fluids to the volume of
urine output, which optimally should be no less than 0.5 ml/kg/h. The animal
usually has a diuresis after the initial dexamethasone dosage and a large portion
of the total i.v. fluids actually comes from the thiopentone.

2.2 Monitoring heart rate and blood pressure

Two measures of the level of anaesthesia are the heart beat and blood pressure.
Monitor the heart rate by shaving a small area on the animal's chest, using
some conductive paste and a small wire which is then amplified and recorded
differentially with the animal's ground (i.e. scalp, stereotaxic attached to animal,
etc.). By using a trigger on the peak of the electrocardiogram (ECG), sequences
of heart beats can be assessed for variability. We prefer, for a run of 20 heart beats
to have a standard deviation of less than 0.05. If the variability becomes larger it
may be a sign of light anaesthesia and hence the need for more thiopentone.
It is important to remember that thiopentone causes myocardial irritability and
may lead to ventricular arrhythmias. These arrhythmias can be treated with
lidocaine i.v., but are best avoided. The blood pressure can be intermittently
checked and lower blood pressures treated with judicious i.v. doses of Isolyte.

2.3 Monitoring end-tidal CO_2

Changes in the end-tidal CO_2 level and especially respiration waveforms (i.e.
voluntary attempts at respiration) provide early indicators of changing
anaesthetic trends. Lower the end-tidal CO_2 to approximately 28–30 mmHg
during placement of the chamber or when the brain surface is open to air. The
intracranial pressure is heavily dependent on the arterial CO_2. Doubling the
arterial CO_2 will double the blood flow, while halving the normal arterial CO_2

will almost halve the blood flow and dramatically decrease swelling. However, during microelectrode recordings and optical recordings the end-tidal CO_2 should be normalized to 36–40 mmHg. End-tidal CO_2 monitoring is useful at the end of the experiment in order to wean the animal from the ventilator (see *Protocol 3*).

2.4 Monitoring EEG

Monitor the animal's EEG with scalp electrodes placed over the cortex opposite to where the chamber is located or over either frontal bones. Differential recordings between scalp and neck muscle will demonstrate EEG changes. These changes are the least reliable indicators of the level of anaesthesia and usually the least sensitive measurement to indicate changes, especially in the presence of nitrous oxide ($>30\%$) which eliminates alpha waves (on which the most certain judgements of anaesthetic depth are based).

2.5 Monitoring the level of paralysis

Monitor the level of paralysis by checking the response of the median or ulnar nerves to stimulation. A baseline twitch response should be checked prior to the initiation of paralysis. The level of paralysis should be kept near to 50%. (At this level, the testing of other responses that are determinants of anaesthetic depth is possible, such as the presence of blink and corneal reflexes.) This level more than adequately suppresses residual eye movements of more than 0.5° (all that is needed for these experiments) and is easily maintained with vecuronium on account of the rapid breakdown of this drug in the liver.

2.6 Brain swelling

When dexamethasone has been given, brain swelling can usually be controlled through the arterial CO_2. This parameter has a direct influence on cerebral blood flow. Doubling arterial P_{CO_2}, for example, will approximately double cerebral blood flow and lead to brain swelling that may result in brain herniation or other detrimental effects. Accordingly, a normal end-tidal CO_2 (36–40 mmHg) is essential during the optical recordings. When the cortex is exposed, the end-tidal CO_2 is reduced to 28–30 mmHg through increased minute ventilation, in order to suppress cerebral blood flow even further. When these measures do not suffice, we have found that a single bolus of thiopentone can be given to combat malignant swelling, presumably by quieting all spontaneous activity which is known to be a major contributor to the cerebral metabolic rate (*note*: a similar treatment, i.e. pentobarbitone coma, is applied to patients with malignant cerebral oedema). Osmotic agents (for example mannitol, urea) are avoided because of the rebound effects and fluid shifts that depress blood pressure and have a generally detrimental effect on optical imaging.

Protocol 3. Recovery

1. After 18–20 hours, conclude the experiment and allow the animal to recover. Discontinue thiopentone and vecuronium.

2. Switch the i.v. solution to one containing 5% dextrose in order to wash out the anaesthetics and muscle relaxants.

3. Hyperventilate the animal to an end-tidal CO_2 of 28–30 mmHg. This will prevent significant brain swelling, but if the brain still appears to be swelling one last dose of thiopentone may be given to decrease overall brain activity.

4. Replace cap for chamber and suture scalp as outlined in Protocol 2, step 8.

5. Allow the end-tidal CO_2 (over 30 minutes) to rise to 38–42 mmHg which will stimulate the animal to breathe on its own. The twitch response can be checked to ensure that a full 100% twitch response has returned. Thus, once the animal has no evidence of active paralysis present, the end-tidal CO_2 demonstrates that the animal should have stimulus to breathe on its own, and the anaesthetic has worn off, the respiratory apparatus can be placed in a bypass mode. (This provides continuous oxygen but bypasses the ventilatory mode.)

6. Gently remove the contacts by irrigating the eye while taking the contacts out. A small amount of dexamethasone and antibiotic ointment are placed over the cornea.

7. As the animal begins to attempt voluntary respirations, give one last dose of antibiotic (ampicillin and chloramphenicol (each 60 mg/kg) and dexamethasone (0.15 mg/kg). The i.v. line can then be removed as can the urinary catheter.

8. When the animal is breathing on its own with no significant increase in the end-tidal CO_2, remove the endotracheal tube and replace the animal in his cage on a heated waterpad. The animal should be up and around in the cage in the next 1–2 hours and should be carefully watched until he is moving around in the cage and able to take food and water.

9. Give injections of antibiotics and dexamethasone i.m. twice per day to prevent infection and brain swelling.

3. Optical imaging: equipment

Imaging hardware is specialized, and there are many options available that probably can be used to visualize neuronal activity. This section covers only the equipment (for example image processor, cameras) and paradigms that we have found useful.

A schematic illustration of the apparatus used to record optical signals appears in *Figure 2a*. Light from a tungsten halogen lamp passes via a fibre-optic sequentially through a condenser, interference filter, and diaphragm, and is then deflected (by either a beam splitter or prism) through an objective lens on to the cortical surface. Reflected light captured by the objective lens is relayed through the beam splitter to a projection lens that focuses it on a photodiode array or a Newvicon TV camera (Cohu 5300). The output of the video camera is fed to the image processor and the final resultant images are displayed on a TV monitor.

3.1 Camera input

The TV camera (Cohu 5300) relies on a Newvicon imaging element, which has the best sensitivity available without an image intensifier (which is undesirable because of the noise it introduces). The Cohu 5300 is available with a signal/noise ratio of 60 dB which is one of the highest available on the market. By operating it at minimum black level, the large DC bias (induced by a brightly illuminated cortex) can be subtracted out, leaving a difference signal consisting almost entirely of variations in surface reflectance. This signal can be amplified greatly, to the point where it fills the entire 1 V peak–peak range of the RS-170 band, so that it can be digitized into 256 separate grey levels by an 8-bit flash A/D convertor (Imaging and Sensing Technology, Inc.) Recent advances in CCD (charge coupled devices) cameras (Photometrics, Hamamatsu) may eliminate the lag time inherent in tube-type cameras.

3.2 Image processor

The digitized output of the flash A/D convertor is fed to an arithmetic processor (ALU-512) which accumulates frames and stores them in one of two $512 \times 480 \times 16$-bit frame buffers (FB-512). Even though each frame is captured with 8 bits of grey level, they are added into a 16 bit buffer. If each incoming frame uses most of the 256 grey levels available to it, the buffer begins to overflow after 256 frames have been added, and overflows several times during the collection of 1800 frames. The overflow could be prevented by using a 24 bit buffer, but it does not really matter. When one image is subtracted from another, the upper bits cancel in any case, and 16 bits suffice so long as the changes of interest lie within them.

The grey level resolution of this system can be enhanced further through lateral averaging, accomplished either by convolving the image with a Gaussian operator or by combining the values of adjacent pixels directly. Just as frame accumulation trades temporal resolution for dynamic range, lateral averaging exchanges spatial resolution for dynamic range. When spatial resolution is reduced (by averaging 4×4 squares of pixels) to 128×120 from 512×480, the dynamic range (signal/noise ratio) can increase by a theoretical maximum of $4 \ (= \sqrt{16})$. The actual improvement is somewhat less, however (Blasdel, unpublished observations).

a

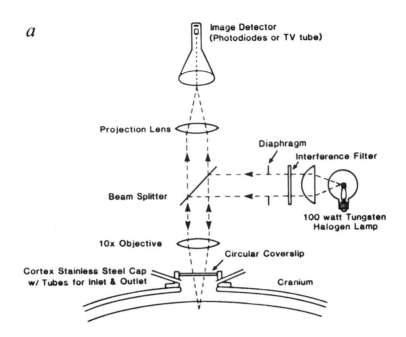

b

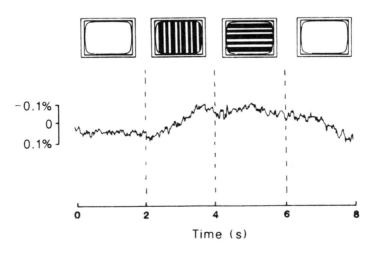

Time (s)

3.3 Subsequent processing

The forgoing procedures can be used to produce images of neural activity which, because they contain noise at many spatial frequencies, can be improved further (at any particular frequency) by filtering. The trick lies in knowing the spatial frequencies that are important. In the case of monkey ocular dominance columns, it is clear from previous work (11) that most of the interesting organizations repeat at intervals of 0.5–1.0 mm. By removing spatial frequencies above and below 1 cycle per millimetre of cortical surface, therefore, one can enhance the dynamic range of signals at this periodicity.

Spatial frequency filtering can best be done by taking a two-dimensional Fourier Transform of the image (moving it from the spatial domain into the frequency domain), zeroing unwanted frequencies, and then moving the information back into the spatial domain. Alternatively, and somewhat more conveniently (with an image processor), one can remove high spatial frequencies by convolving with a narrow Gaussian operator, and low spatial frequencies by subtracting out the result of convolving with a wide Gaussian operator.

4. Optical imaging: biology

Optical signals may be obtained with or without optical probes of membrane potential (voltage-sensitive dyes). The signals and advantages associated with each approach are described below.

4.1 Voltage-sensitive dyes

Blasdel and Salama (7) screened many voltage-sensitive dyes, for use in monkey striate cortex, and found that one (NK2367, Nippon Kankoh–Shikiso Kenykyusho Co., Ltd) consistently produced the largest signals. The reasons for this are still unclear, but probably result from the rapid penetration of this dye into the upper layers of cortex, and the rapid bleaching of it from the pia with the application of light. More stable probes (for example RH155, Molecular Probes), that penetrate the cortex poorly, give much larger signals when the dye is transilluminated and do not work nearly as well as NK2367, presumably due to light blockage by darkly stained cuboidal cells in the pia.

Figure 2. Imaging apparatus: (a) Lower portion of *Figure 2a* is the monkey skull with a 25 mm diameter window in the cranium and stainless steel chamber with inlet/outlet ports for perfusing dye over cortex (after it is sealed with insert having glass coverslip). Collimated light from a tungsten halogen lamp driven by constant DC power supply passes through an interference filter (720 ± 20 nm), aperture, and is reflected to illuminate the cortical surface with a beam splitter. Reflected and scattered light from the cortical surface passes through objective and projection lenses to a photodiode array or video camera. (b) Average responses to 10 presentations of either blank screen, vertical or horizontal gratings in a 100 µm² patch of primary visual cortex viewed by single photodiode. Baseline is established while the monkey views a blank screen. An abrupt change in absorption occurs when the vertical grating appears. A small deflection occurs when the presentation changes from vertical to horizontal gratings and the signal returns to baseline when the blank screen reappears. (From Blasdel and Salama, 1986, with permission.)

Since the optical signals obtained with NK2367 fade over time and return with reapplication of the dye, it is likely they derive, at least in part, from the dye. The likely sources of these dye-dependent components are: dye-stained neurons and dye-stained glia, which are considered separately below.

4.2 Dye-stained neurons

Fast, dye-dependent signals probably derive from the depolarization of neural membranes, most of which are contained in fine axons and dendrites in the neuropil. Photodiode arrays with a temporal resolution of less than 1 ms can detect voltage-sensitive dye signals. Even though they occur quickly, within milliseconds (or less), these depolarizations can also be detected by slow devices (for example television cameras) due to changes in the time-integrated potential.

4.3 Dye-stained glia

Slower and larger signals are expected from glial cells responding secondarily to the accumulation of potassium. Intracellular recordings in cat striate cortex (12) have shown that glial cells depolarize by 5 mV or more in response to visual stimulation, and that they can be just as orientation-selective as neurons. Since glial cells act like potassium ion-sensitive microelectrodes, neuronal activity in a specific region will cause local increases in extracellular potassium and glial depolarization. Optical recordings from presumed glial cells, moreover (9), show a continuous activity-driven change in absorption.

4.4 Practical aspects of dye application

When dyes are used to obtain recordings of neural activity, one must choose a dye (a) that produces the largest possible signal, (b) that produces the fewest toxic effects, and (c) that penetrates into the tissue. In the case of primate cortex, the third criterion is the hardest to satisfy since the pia mater presents a barrier to dye penetration. Some voltage-sensitive dyes that work well in tissue slices (for example, RH-155) and in heart muscle have limited uses in cortex because they cannot pass through the pia.

There are two basic ways of circumventing this problem. One is to find ways of inducing transport across the pial surface. The other is to find dyes that are more permeant. While the optimal balance between these two approaches is not yet clear, there are many problems with the first one. Attempts at penetrating the pial barrier (by injecting dye subpially through a glass micropipette) or disrupting it (through osmotic shock), usually leads to an unresponsive cortex, or to excessive bleeding or platelet aggregation that obscures the cortical surface. Accordingly, the second approach appears preferable. There are many voltage-sensitive dyes available; one has merely to find a good one that penetrates the tissue. One of the main advantages of NK-2367 lies in its high diffusibility which permits it to cross the pia.

4.5 Intrinsic signals

Visual stimuli cause the cortex to darken, even before it is stained (7). Grinvald *et al.* (13) have studied the phenomenon in greater detail and coined the term 'intrinsic signal' to describe it. Frostig (10) and Ts'o (14) and their colleagues showed that intrinsic signals may also be used to visualize patterns of cortical activation. Even though the mechanisms responsible for intrinsic signals are not well-understood, they could derive from a number of different mechanisms, including the dilation of capillary beds and the increased scattering of light from the activity-dependent release of potassium, or from the swelling of neurons and/or glial cells (15). These signals offer an attractive alternative to optical probes in the study of cortical events where such probes are ill-advised or impractical.

The major drawback to intrinsic signals is that they are slow. They require 1–1.5 seconds to begin (in monkey striate cortex), and several more to mature (13). As one can see from the photodiode trace in *Figure 2b*, signals from dye-stained cortex appear more quickly (with an onset latency of 200 ms and a maturation within 1.5 s), so consequently are easier to isolate from visually-induced changes in the vasculature (which also responds with a time course of 2–3 s). A second drawback is their likely involvement with vascular sources. If these include laterally projecting blood vessels, the intrinsic signals can give erroneous information, indicating activity in laterally displaced areas, for example, where there is none. Also, since intrinsic signals summate over larger distances than those obtained with voltage-sensitive dyes (which block light and consequently restrict its penetration to the upper layers), there is greater ambiguity about the depth of monitored activity; this can be a problem if response properties are not identical through depth (as they rarely are). Geometrical constraints (for example, parallax—see below) also pose a greater problem, because the optical axis is rarely aligned with that of cortical 'columns'.

Intrinsic signals do provide several important advantages, however. Because staining is not required, they can be exploited to monitor activity through the intact dura, and even through bone if it is thin enough (10). Illumination without dyes also avoids the toxic effects associated with photo-oxidation, rendering this approach much more suitable to clinical applications.

4.6 Vascular artefacts

Grinvald *et al.* (13) noted that some blood vessels change their absorption in response to visual stimulation, and suggested that these changes might contribute to the images reported by Blasdel and Salama (7). Such artefacts are less of a problem with voltage-sensitive dyes than with intrinsic signals, however, because the rapid dynamics of dye-related signals allow them to be isolated more easily (16). The high resolution of video imaging makes it obvious, moreover, when vascular artefacts occur because they *look* like blood vessels. Their contributions can then be mitigated through the adjustment of experimental parameters, the most important of which is timing. Even though blood vessels respond to visual stimuli they do so more slowly than neurons, and the difference in latency can be exploited to minimize (or remove) their contribution.

4.7 Geometric constraints

Mammalian cortex is relatively transparent, raising several issues concerning alignment. If activities are averaged simultaneously through a millimetre or more of cortex, and response properties are not aligned precisely along the path of averaging (in this case the optical axis), the consequences can be severe. A divergence of only 15° for example, induces a registration error of 260 μm— half the width of an ocular dominance band—which reduces contrast. It can also cause distortion through parallax.

It is difficult to avoid this problem because response properties are not always aligned perpendicular to the cortical surface. The axis along which they are aligned (the one defined by axon bundles and apical dendrites) varies, and is difficult to judge *in vivo*. The problems can be minimized, however, by focusing on activity in the upper layers. This is one of the indirect benefits of using light-absorbing dyes. By limiting the depth of light penetration in the cortex, they ensure that all optical signals, even intrinsic ones, come preferentially from the upper layers.

4.8 Noise

Sources of noise that prevent signal detection include: (a) neural, (b) electrical, and (c) vascular noise. Noise from stained neurons firing stochastically, is inherent to any healthy neuronal system and must be dealt with through averaging. Electrical noise can be minimized through the selection of equipment which should, in any case, be able to detect 0.01% changes in radiance. Vascular noise derives from the tight coupling of the microcirculation and neuronal activity. Even when the cortex looks stable (through an operating microscope, for example), a photo-diode focused on a minimally vascular part of cortex still records heart beat and respiration—due, presumably, to blood corpuscles squeezing their way through capillaries—and these signals are at least as large, if not larger, than those associated with neural activity.

Our approach to minimizing vascular noise has been, first, to seal the system hydraulically. Until this is accomplished there is no hope of detecting optical signals induced by activity. Viewed through an operating microscope or video camera, the cortex should appear absolutely stable. Once accomplished, one should ensure that the animal is adequately and constantly anaesthetized so that the heart beat is regular and predictable. Finally, frame collection should be synchronized to the ECG and to the respiration so that image and control frames will be disturbed to roughly comparable extents by fluctuations in blood pressure and cerebral blood flow.

Photodynamic damage results from interaction between dye molecules, light, and neurons. Needless to say, the effects on neural tissue are usually adverse. Photobleaching results directly from exposure of the dye to light. While it is essential that some photobleaching occurs immediately after the cortex has been stained, to remove dye from the pia which will have stained quite darkly, photobleaching can be a problem if dye fades so quickly and unevenly that its distribution changes markedly during the accumulation of frames. Both problems (photobleaching and photodynamic damage) are alleviated by using the lowest possible amount of illumination, making it imperative to use the highest possible numerical aperture for light collection (where numerical aperture is defined as the diameter of the lens diaphragm divided by twice the distance between the lens and the tissue). A numerical aperture greater than 0.1 is highly desirable, although the upper limit is determined by the depth of field required. Recent studies have used very high numerical aperture lenses in an attempt to narrow the depth of field and allow focusing at different depths (17).

5. Optical imaging: applications

The changes in optical absorption induced by visual stimulation are understood most easily by considering the optical signals generated by a small patch of cortex, 100 µm across (*Figure 2b*). During the first 2 second interval, one can see a baseline established while the animal views a blank screen. The signal undergoes a sharp inflection when the first visual stimulus, a drifting vertical grating, appears, and then rises steadily. A small inflection occurs when the orientation of contours switches from vertical to horizontal. The signal decays back to baseline when the horizontal contours disappear.

The large size and long duration for which these signals can be sustained makes them detectable by a television camera, with higher resolution. The differential imaging strategy developed to obtain data can be understood by considering two images of cortex acquired at strategic intervals. One is averaged for 1.5 s at the end of the first period, while the animal views a blank grey screen. The other is obtained at the end of the second period, while the animal (in this example) views a drifting vertical grating. By subtracting the first image from the second, one is left with a third, difference image of regions where the optical absorption has changed in response to visual stimulation. By repeating the sequence several times and averaging the results, one can diminish contributions from fluctuations that do not occur in synchrony with visual stimulation.

If video frames are collected separately, during stimulation of the right and left eyes, images of ocular dominance bands are produced. If collected while contours are presented at two orthogonal orientations, orientation patterns are produced. By analogy, if frames are collected while electrical stimuli are, and are not, applied to a stimulating electrode, an image of the regions activated by this simulation (those receiving input from cells around the electrode) is produced. Examples with specific protocols are shown in Sections 5.1 and 5.2.

Four examples of video imaging are shown in this section. Two deal with differential images of ocular dominance and orientation preference in monkey striate cortex; further details of which have already been described in detail (14, 16).

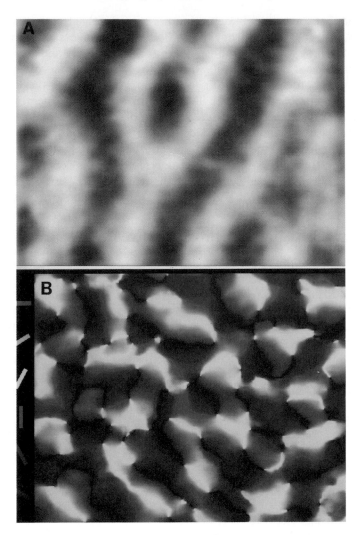

Plate 1. Voltage-sensitive dye images of ocular dominance columns and orientation preferences in primary visual cortex. Video images using a voltage-sensitive dye (NK2367) obtained from a 4 mm² patch of area 17 with the lowermost portion of (A) near and parallel to the 17/18 border. An average of frames from stimulation of the left eye is subtracted from averages of video images from the right eye (for example, differential video imaging). (A) Averaged frames from stimulation of the left eye are subtracted from those averaged during stimulation of the right eye (all orientations) resulting in ocular dominance bands (right = black; left = white). The bands are approximately 500 μm wide, consistent with DG and electrophysiological studies. (B) The same patch of cortex but differential video imaging of both eyes stimulated at orthogonal orientations (for example, vertical–horizontal; left oblique–right oblique). This figure represents a summary of areas of cortical selectivity for specific orientations. Note that with the voltage-sensitive dye there is little or no vascular artefact.

The other two deal with preliminary efforts (one *in vitro* in monkey visual cortex tissue slices, the other *in vivo* in the intact cortex of an adult monkey) to trace neuronal pathways through differential video imaging.

5.1 Ocular dominance and orientation

Plate 1A/B show an example of differentially imaged ocular dominance and orientation patterns, from one region of cortex (4 mm²), obtained using the voltage-sensitive dye, NK2367. The ocular dominance pattern (*Plate 1A*) was achieved by alternately stimulating the two eyes with contours at all orientations while two sets of frames were collected. The orientation pattern (*Plate 1B*) was achieved by stimulating the animal alternately with contours at orthogonal orientations. As one can see, the resulting patterns are unambiguous.

Protocol 4. Imaging ocular dominance columns

1. Using microelectrode recordings, identify a single unit and its receptive field. Align the eyes using a small mirror just in front of one eye.

2. With the eyes aligned, make 3–4 vertical microelectrode penetrations and photograph each site for later comparison with video images. Each cell should be characterized for ocular dominance, orientation preference, and location of receptive field.

3. Remove the microelectrode recording insert and replace it with the video imaging insert (with glass coverslip inserted). The video imaging insert has two ports, one for infusion of dye and one for drainage.

4. Mix a small quantity of the voltage-sensitive dye (NK-2367) to give a 1% solution (15 ml) in normal saline or balanced salt solution (BSS). The saline or BSS should be warmed by microwave prior to mixing in dye to remove bubbles. The outlet tubing should be kept no more than 5 cm above the top of the head, secondary to risk of increased intracranial pressure. Perfuse the dye solution over the cortex for 1–2 h, then wash out for 5–10 min. After the first 5–10 ml is perfused, the dye may be recirculated.

5. Seal the chamber by turning off the pump for infusion and letting fluid drip from outlet tubing which is then sealed (3-way stopcock).

6. Using green light, focus the imaging microscope on the blood vessels and store the photo for later identification of images taken and for comparison with microelectrode recording sites.

7. Turn the gain on the camera (COHU 5300) to maximum and the white level to minimum. Under red light (720 nm) with a constant DC power supply, turn the light up until it just fills the field (and just below saturating the camera).

8. Use a shutter system to keep light off except during frame capture.

9. As the moving grid patterns appear on the TV screen to stimulate the monkey, heart beat and respiration are synchronized.

Protocol 4. *Continued*

10. Cover the right eye with the shutter and stimulate with the moving grating for 3 s (180 frames). These frames are stored in one frame buffer.

11. Then after a 3–6 s delay, cover the left eye and collect the right eye frames (180 frames).

12. After repeating this sequence 4–6 times, subtract the averaged left from the right eye image, leaving a differential image (*Plate 1A*) where black represents the right and white represents the left eye.

Problems

[a]If the level of anaesthesia becomes too light in the middle of the run, blood vessel artefacts will appear. Check the level of anaesthesia (see Section 2.5).

[b]Occasionally on the first few runs the dye-staining of the pia is either not bleached enough or the initial bleaching of the dye leads to a circular pattern of activity. Repeat the run.

[c]If initially imaging good quality ocular dominance columns which then fade, check that level of anaesthesia/paralysis is not too deep or reapply dye for 20–30 minutes (see Step 4).

[d]If the signals are consistently weak, check blood pressure, fluid balance, and level of anaesthesia to rule out the possibility of consistent low blood pressure or too much anaesthesia as the cause.

Protocol 5. Imaging orientation preferences

1. See *Protocol 4*, steps 1–9, with special emphasis on close eye alignment during the microelectrode recordings.

2. Show moving gratings of orthogonal orientations to the animal (i.e. first 0° then 90°).

- Open the light shutter while moving vertical bars appear on TV screen, and store 3 seconds (180 frames) of images.

- Then after a variable delay (usually 3–5 s), store 3 seconds of images from the orthogonal orientation of moving grating (in this case, horizontal).

3. Subtract the vertical image from the horizontal image to show patches of cortex which prefer horizontal orientation stimulation.

4. Repeat this paradigm for a number of angles (0, 22.5, 45, 67.5, 90, 112.5, 135, 157.5, 180) (*see Figure 3* for raw data appearance).

5. Combine the images by checking, pixel by pixel, which orientation has the strongest signal and assigning a specific colour for each orientation (*Plate 1B*).

In *Plate 2A/B*, one can see comparable examples of ocular dominance and orientation (from a different animal), obtained in the absence of voltage-sensitive dyes (i.e. with intrinsic signals). These patterns were obtained through differential imaging, although the period of visual stimulation necessary was longer, and more frames were averaged (to improve signal to noise ratio). As pointed out

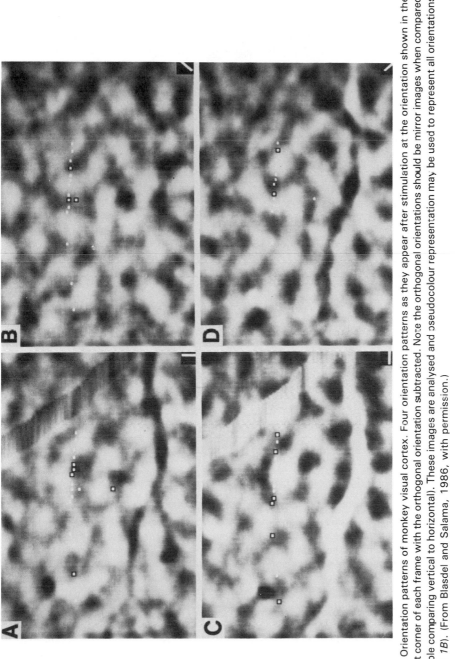

Figure 3. Orientation patterns of monkey visual cortex. Four orientation patterns as they appear after stimulation at the orientation shown in the lower right corner of each frame with the orthogonal orientation subtracted. Note the orthogonal orientations should be mirror images when compared (for example comparing vertical to horizontal). These images are analysed and pseudocolour representation may be used to represent all orientations (see *Plate 1B*). (From Blasdel and Salama, 1986, with permission.)

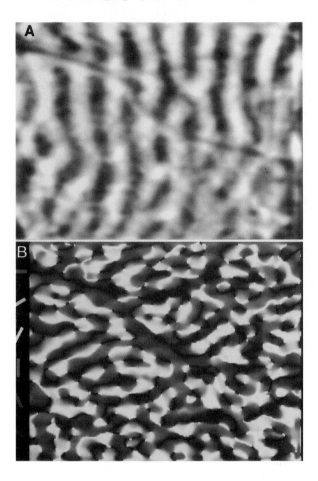

Plate 2. Intrinsic signal images of visual cortex. Ocular dominance columns and orientation preference maps using the intrinsic signal in a 8 mm^2 patch of visual cortex. As in *Plate 1*, the lower border of the figure is parallel to the area 17/18 borders. (A) Ocular dominance columns demonstrated by right–left eye stimulation with moving gratings at all orientations. The ocular dominance columns are approximately 500 µm wide. However, with the intrinsic signal imaging, in order to get large enough S/N ratios, more frames are collected (15%) and more vascular artefacts are apparent. (B) Orientation preference summary map. Again note the moderate sized vascular artefact. The patterns are consistent with those collected with the dye.

earlier, these patterns probably reflect activity averaged through a greater depth of cortex. Furthermore, because of the longer time constants (approximating those of blood vessels), vascular artefacts are more apparent. This drawback is outweighed, however, by the unquestioned advantage of not having to use dye in some (for example clinical) applications.

5.2 *In vivo* stimulation studies

Most early work on the microcircuitry of monkey striate cortex has been derived from anatomical studies. Some of the most detailed anatomy has been provided at the single cell level by the Golgi stain (18, 19), and through intracellular recording from and filling cells with horseradish peroxidase (HRP) (20). Pathways involving larger numbers of cells have been studied successfully, following injection and transport of various tracer substances (21–23). A major problem with these approaches, however, has always been that of linking the observed projections to patterns of physiological activity (2). An example of how this problem might be overcome is provided by differential imaging. Indeed, differential imaging of cortical projections may provide an ideal method for analysing cortical projections from one layer to another (in tissue slices), as well as from one laterally displaced region to another (in the intact cortex), while maintaining functional integrity so that pathways can be investigated interactively in the context of other projections and observed physiological activity.

Lateral projections in monkey visual cortex can be studied *in vivo* with video imaging paradigms. The main difference is that the animal remains intact and the more rapidly oxidized dye NK2367 is used. Because the signals obtained with this dye are also much smaller, ten times as many averages (300) are required before and after the stimulus is applied, compared to a total of 30 frames before and after stimulation in the brain slice preparation.

Protocol 6. *In vivo* stimulation studies

1. Using the apparatus for the microelectrode recording, place a bipolar stimulating electrode (30-gauge wire coated with insulation except at the two tips—5–20 µm back and 2–5 µm in diameter at tip) 200–100 µm below the surface of the cortex.

2. Set-up video imaging apparatus as detailed in *Protocol 4*.

3. Synchronize heartbeat and respiration just prior to collection of control images.

4. Collect 30 frames (500 ms) before stimulation and store in one frame buffer (control images).

5. Stimulate at 10 Hz (biphasic pulse, peak-to-peak 0.05–0.4 mA amplitude) for 1–3 seconds.

6. Collect 30 frames after stimulation and store in a separate frame buffer (stimulation images).

7. Wait 10–30 s between stimulations and collect 10 trials averaged in their respective frame buffers, control and stimulation.

8. Subtract the overall average of the control images from the stimulation image (*see Figure 4*).

Protocol 6. *Continued*

9. With the stimulation intensity at a very low level repeat sequence above. The resultant image should show no activity (*Figure 4A*). Multiple parameters can be adjusted including number of frames collected, stimulation depth, intensity, duration, and frequency.

Figure 4 shows an example of results obtained with a stimulating electrode placed 500 μm beneath the cortical surface. *Figure 4A* shows the control, where the stimulation intensity was maintained well below threshold. The resultant image shows the lack of activity which should be obtained for the control situation. When the stimulation intensity was raised above threshold, repeatable and consistent patterns were observed (*Figure 4B*). The larger, patchy dark area, vertically above the tip of the stimulating electrode, probably derives from the depolarization of apical dendrites and other vertically-projecting processes (19, 22), while the alternating light and dark stripes in the lower left hand corner probably reflect the activity of periodically distributed lateral projections (23).

5.3 Visual cortex slice stimulation

An example of video imaging in tissue slices from monkey striate cortex appears in Figure 5. The details of the slice preparation are beyond the scope of this chapter (24). As an example of the strength of the video imaging technique, the same protocol was used as for *in vivo* imaging. Because the better voltage-sensitive dyes for brain slices (for example RH-155) give larger signals and light can be applied in a transmission mode, frame averaging can be limited to single trials.

A piece of cortical tissue was surgically removed from striate cortex in an anaesthesized animal, immediately prior to sacrificing the animal. The tissue was taken from an area just behind the border between areas 17 and 18. The slices were cut parallel to the 17/18 border, so that functional ocular dominance slabs run into the surface (3). Bipolar stimulation (see *Protocol 6*) was performed near layer 4 for 1 s at 10 Hz. Video images of the voltage-sensitive dye activity (RH-155) were obtained for the 500 ms (30 frames) before and after stimulation. The average obtained before stimulation was then subtracted from that obtained after stimulation. In these slice experiments only one trial was necessary to get excellent images, whereas during *in vivo* imaging using similar stimulation paradigms, 10 trials were necessary. A control image appears in *Figure 5A*, where the stimulation current was less than 0.05 mA. When the intensity was increased to 0.3 mA, a consistent pattern developed (shown in *Figure 5B*). By increasing the stimulation intensity to 0.4 mA, there was a significant and consistent increase in the slice activity (*Figure 5C/D*). The size of the image is approximately 2 mm². The width of a stripe is from 125–150 μm. These studies are only preliminary and the anatomical substrate of these bands is currently under study.

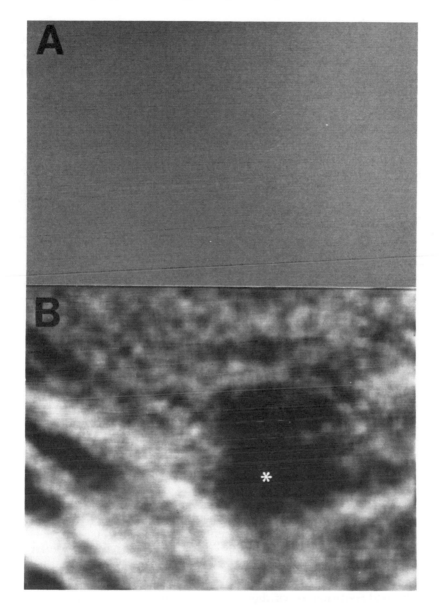

Figure 4. Visual cortex stimulation *in vivo*. *In vivo* stimulation of a 4 mm² patch of visual cortex. The stimulating electrode was placed approximately 500 µm directly below the (✿) in (B). Differential video images were collected by subtracting pre- from post-stimulus averages (30 frames × 10 trials). Stimulation was at 10 Hz, bipolar, biphasic, at either 0.5 mA or 10 V. (A) Control images with subthreshold stimulation intensity (1 V) shows no activation. (B) When the stimulation intensity is increased to 10 V, activation of areas directly above the stimulating electrode occur as well as a striped pattern of alternating depolarization/hyperpolarization (black/white) with widths on the order of 200–300 µm.

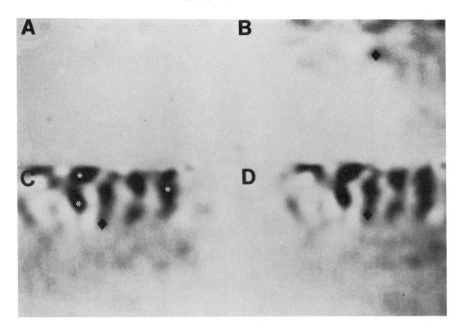

Figure 5. Visual cortex slice stimulation. The images are 2 mm² with the pial surface at the top and the white matter at the bottom. Each of the four images are single trials of 50 msec (30 frames) before stimulation subtracted from the 500 msec immediately following stimulation (bipolar, biphasic, 10 Hz, 1 s). The site of stimulation is the black diamond (*Figure 5B,C,D*). (A) Low level stimulation (0.05 mA) produces no significant activation. (B) Increasing the stimulation intensity to 0.3 mA produces a patchy distribution of activation (black). (C) Increasing the stimulation intensity to just 0.4 mA causes dramatic activation of a striped nature with alternating areas of activation (black) and hyperpolarization (white). During single shock stimulations, extracellular recordings (∗) demonstrated field EPSPs and population spikes. (D) Same intensity stimulation as (C), but 20 minutes later, shows consistent pattern of activation. The stripes are approximately 125–150 μm wide.

When extracellular microelectrode recordings were made from the black (presumed excitatory) regions represented in *Figure 4C* (sites are labelled with ∗), field EPSPs and population spikes were observed. However, when extracellular microelectrode recordings were made from the white (presumed inhibitory) areas, no extracellular field potentials or population spikes were found. Final confirmation of these 'inhibitory' areas will require intracellular recordings.

6. Summary

The above examples of differential imaging with video cameras were not meant to provide an in-depth review of existing studies so much as to convey a flavour of the possibilities that exist. Whether the investigations are centred on single cultured cells, or on the intact cortex of non-human primates, video imaging is developing into a fast and reliable technique, with unsurpassed resolution.

Acknowledgements

MMH is supported by a Grass Foundation Morison Fellowship and an American Association of Neurological Surgeons Research Foundation Fellowship. GGB is supported by the McKnight Foundation, Office of Naval Research and NIH EY21008.

References

1. Hill, D. K. and Keynes, R. D. (1949). *J. Physiol., Lond.*, **108**, 278.
2. Hubel, D. H. and Wiesel, T. N. (1972). *J. Comp. Neurol.*, **158**, 267.
3. LeVay, S., Hubel, D. H., and Wiesel, T. N. (1975). *J. Comp. Neurol.*, **159**, 559.
4. Sokoloff, L. (1977). *J. Neurochem.*, **29**, 13.
5. Bellugi, U., Poizner, H., and Kilma, E. S. (1989). *Trends Neurosci.*, **12**, 380.
6. Petersen, S. E., Fox, P. T., Snyder, A. Z., and Raichle, M. E. (1990). *Science*, **249**, 1041.
7. Blasdel, G. G. and Salama, G. (1986). *Nature*, **321**, 579.
8. Grinvald, A. (1984). *Annu. Rev. Neurosci.*, **8**, 263.
9. Grinvald, A., Frostig, R. D., Licke, E. E., and Hildasheim, R. (1988). *Physiol. Rev.*, **68**, 1285.
10. Frostig, R. D., Lieke, E. E., Ts'o, D. Y., and Grinvald, A. (1990). *Proc. Natl Acad. Sci.*, **87**, 6082.
11. Hubel, D. H. and Wiesel, T. N. (1977). *Ferrier Lecture, Proc. R. Soc. Lond.*, **198**, 1.
12. Kelly, J. P. and Van Essen, D. C. (1974). Cell structure and function in the visual cortex of the cat. *J. Physiol.*, **238**, 515.
13. Grinvald, A., Lieke, E., Frostig, R. D., Gilbert, C. D., and Wiesel, T. N. (1986). *Nature*, **324**, 361.
14. Ts'o, D. Y., Frostig, R. D., Lieke, E. E., and Grinvald, A. (1990). *Science*, **249**, 417.
15. Hochman, D. and MacVicar, B. A. (1990). *Neurosci. Abstr.*, **16**, 1096.
16. Blasdel, G. G. (1989). *Sensory processing in the mammalian brain* (ed. J. S. Lund) pp. 212–268. Oxford, New York.
17. Malonek, D., Shoham, D., Ratzlaff, E., and Grinvald, A. (1990). *Neurosci. Abstr.*, **16**, 292.
18. Lund, J. S. (1973). *J. Comp. Neurol.*, **147**, 455.
19. Lund, J. S. and Boothe, R. G. (1975). *J. Comp. Neurol.*, **159**, 305.
20. Gilbert, C. D. and Wiesel, T. N. (1979). *Nature*, **280**, 120.
21. Blasdel, G. G. and Fitzpatrick, D. (1984). *J. Neurosci.*, **4**, 880.
22. Blasdel, G. G., Lund, J. S., and Fitzpatrick, D. (1985). *J. Neurosci.*, **5**, 3350.
23. Rockland, K. S. and Lund, J. S. (1982). *Science*, **215**, 1532.
24. Appendix. (1984). In *Brain slices* (ed. R. Dingledine), pp. 381. Plenum Press, New York.

5

In vivo voltammetric methods for monitoring monoamine release and metabolism

J. A. STAMFORD, F. CRESPI, and C. A. MARSDEN

1. Introduction

1.1 What is voltammetry?

Electrochemical detection (ECD), coupled with high performance liquid chromatography (HPLC), has become the method of choice for the assay of trace amounts of amine neurotransmitters and their metabolites in brain tissue samples since its introduction in the seventies (1–3). *In vivo* voltammetry essentially involves implanting a miniaturized version of the HPLC electrochemical detector into the brain.

Electrochemical detection of amines and their metabolites exploits a well-known property of catechol- and indole-based substances, their susceptibility to oxidation (*Figure 1*). The ability of compounds, such as oxygen, to act as oxidizing agents for this reaction relies on the fact that the energies of their unfilled electron orbitals lie sufficiently below those of the filled orbitals of the oxidizable moieties of the catechols or indoles that electron transfer can occur with a net decrease in free energy. The electrochemical detection system embodied in voltammetry uses, as oxidizing agent, a carbon-based electrode to which is applied a positive potential sufficient to remove electrons from the oxidizable compound. The resultant flow of electrons can be measured in the form of an electrical current which is directly proportional to the amount of material oxidized.

The major difference between the use of electrochemical detection for HPLC and *in vivo* voltammetry is the absence of a chromatographic separation procedure in the latter which, instead, relies upon the individual oxidation potentials of the compounds of interest to provide resolution. Some electroactive species undergo oxidation with greater ease than others, and by slowly increasing the applied potential individual substrates in the extracellular space can be sequentially oxidized. For instance, dopamine (DA) will oxidize at a lower potential than tyrosine since the latter has only one hydroxyl group and is,

In vivo *voltammetry*

Redox equations

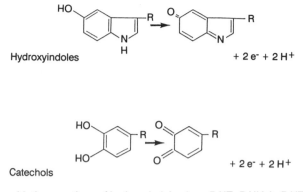

Hydroxyindoles $+ 2e^- + 2H^+$

Catechols $+ 2e^- + 2H^+$

Figure 1. The oxidation reactions of hydroxyindoles (e.g. 5-HT, 5-HIAA, 5-HTP) and catechols (e.g. DA, NA, DOPAC) result in the production of current ($2e^-$).

therefore, more difficult to oxidize. Similarly, tryptophan oxidizes at a higher potential than 5-hydroxytryptamine (5-HT) as its oxidation centres upon the NH-group in the indole ring, while 5-HT also has the more readily oxidizable hydroxyl group. Catechols are easier to oxidize than indoles since electrons are more readily removed from hydroxyl groups than from NH-groups. Metabolic deamination of catechols and indoles has little effect on their oxidation potentials: the metabolites, dihydroxyphenylacetic acid (DOPAC) and 5-hydroxyindoleacetic acid (5-HIAA) oxidize at the same potentials as their parent amines, DA and 5-HT. This is a key problem in the interpretation of *in vivo* signals and is discussed later.

Although there are numerous chemical compounds in the brain the *in vivo* electrochemist is fortunate that relatively few are both present in high enough concentrations in the extracellular fluid to be detected, and have oxidation potentials within the usable potential range. This 'potential window' extends from approximately -0.2 to $+0.8$ V versus a Ag/AgCl reference with slow scanning methods (although faster techniques have a wider range). At potentials more negative than -0.2 V oxygen is reduced, of which there is an ample supply in the brain, resulting in the generation of excessive current overshadowing other redox processes. Similarly, the upper potential limit is determined by the oxidation of water. *Table 1* shows the approximate redox potentials of compounds of neurochemical interest oxidizable within this potential window. Ranges (rather than specific values) are given since the precise potential will vary with the type of working electrode and method of voltammetric measurement (Section 1.2). Several drugs used to manipulate catechol and indole neurotransmission are electroactive (oxidizable) and this must also be considered when designing experiments. Ascorbic acid and uric acid, not readily associated with neurotransmission, are included in the table as they are present

114

Table 1. Approximate[a] oxidation potentials of compounds of neurochemical interest at carbon-based electrodes *in vitro*

Compound	Potential vs Ag/AgCl
Ascorbic acid	−0.2 to +0.2
Dopamine Dihydroxyphenylalanine Dihydroxyphenylacetic acid Noradrenaline Adrenaline	+0.15 to +0.25
5-hydroxytryptophan 5-hydroxytryptamine 5 hydroxindole acetic acid	+0.3 to +0.40
Uric acid	+0.3 to +0.40
3-methoxytyramine Homovanillic acid Normetanephrine	+0.4 to +0.45
Tyramine Tryptamine Octopamine Tryptophan Tyrosine Cysteine Neuropeptides containing tyrosine or tryptophan	+0.75 to +0.90

[a] The exact oxidation potentials are not given as these will vary *in vitro* with changes in temperature, pH, and the type of carbon electrode (see Section 1.3.2). The list is not comprehensive and it is important to test the electroactivity of any drugs administered during *in vivo* work. The oxidation potential measured *in vitro* does not imply that the compound will oxidize at **exactly** the same potential *in vivo*.

in high concentrations in brain extracellular fluid (4) and cause particular problems since they have oxidation potentials very similar to catechols and indoles, respectively.

The degree to which individual oxidizable species can be resolved in the brain with *in vivo* voltammetry is largely dependent upon two factors: the type of electrode employed and the measurement technique (the way in which the potential is applied to the electrode).

1.2 Introduction to voltammetric measurement techniques

All voltammetric techniques are concerned with measuring the current produced by oxidation or reduction of chemicals following the application of a suitable potential. This redox reaction occurs at the surface of the working electrode to which the chemical in solution (*in vitro*) or in the extracellular fluid (*in vivo*) moves by diffusion. In most voltammetric methods, the applied voltage is not maintained at a constant value, but is instead a waveform (for example a ramp,

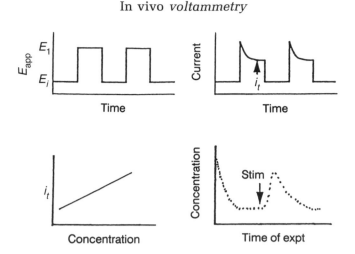

Figure 2. The basic principles of chronoamperometry. **Top left:** The technique consists of the repetitive application of pulses of voltage (E_{app}) from an initial resting potential (E_i) to a value sufficient to oxidize the compound(s) of interest (E_1). **Top right:** Current output from an electrode. The current (i) is measured at a fixed time (t) towards the end of the pulse after the charging current has decayed. **Bottom left:** The sampled oxidation current (i_t) is proportional to the concentration of electroactive species. **Bottom right:** In a typical chronoamperometric experiment the concentration of electroactive species are plotted as a function of time. Once a stable baseline has been obtained a stimulus (stim) may be given.

cyclic ramp, series of pulses). The voltage input waveform determines the faradaic to charging current ratio, the first current due to the oxidation (and reduction) of the species at the active surface of the working electrode, the second due to charging of the electrical double layer of the electrode surface. In general the slower the voltage scan rate, the higher the faradaic current to charging current ratio, the faradaic current being proportional to the square root of the scan rate, while the charging current is directly proportional to the scan rate (5).

The simplest measurement technique is *chronoamperometry* where the voltage is increased instantaneously to a potential sufficient to oxidize the electroactive compounds and held for a fixed time (50 ms to 1 s). The current shows a sharp transient rise followed by decay to a steady level and is measured immediately before the end of the pulse. This gives an accurate evaluation of the concentration of electroactive species but with poor selectivity, since all species oxidizable at a potential equal to or less than the applied potential will contribute to the oxidation current (*Figure 2*).

Another approach is to use a linearly increasing voltage ramp. Faradaic current does not occur until the voltage reaches the oxidation potential of the compounds in solution (or present in the extracellular fluid). At these voltages the current rises to a plateau (or a peak at high voltage scan rates). Thus, if two electroactive species oxidize at sufficiently different potentials, two distinct plateaux (or peaks) occur. This technique is called *linear sweep voltammetry*. It can provide

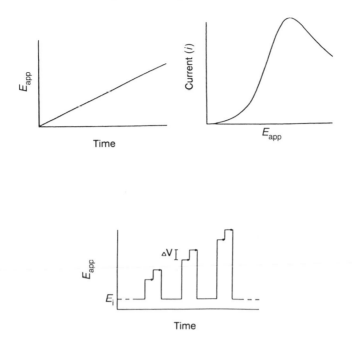

Figure 3. **Top:** Linear sweep voltammetry. **Left:** The input waveform consists of a linearly increasing applied potential (E_{app}) versus time. **Right:** The current (i) output from the electrode reaches a peak versus the input voltage (E_{app}) as the compound of interest is oxidized. **Bottom:** Differential double pulse voltammetry. The applied voltage (E_{app}) consists of a double potential step consisting of a series of sequentially increasing voltage pulses upon which are superimposed a second fixed voltage increment (ΔV). Current is sampled on each pulse immediately before and towards the end of the fixed voltage increment. The voltage is held at a resting potential (E_i) between steps.

information on the concentration of electroactive species together with, in some cases, identification of the electroactive compounds (*Figure 3*). An extension of this technique is *cyclic voltammetry*. Here, a reduction sweep follows the oxidation sweep, such that compounds oxidized on the initial scan can be reduced on the reverse scan providing further characterization of the chemical identity of the oxidizable compound (1).

More recently cyclic voltammetry has been modified for use with high scan rates (300 V/s) with each scan lasting 10–20 ms; a technique termed *fast cyclic voltammetry* (FCV). Increasing the scan rate causes less disruption to the brain environment. When the electroactive compound is adsorbed on to the electrode surface it undergoes a cycle of oxidation and reduction which is manifested as incremental currents superimposed on the background current (Section 2.2).

The linear sweep technique can be modified to produce peak-shaped voltammograms by applying a linear potential ramp (5–10 mV/s) with superimposed (3–5 times/s) regular step potentials of constant amplitude and

117

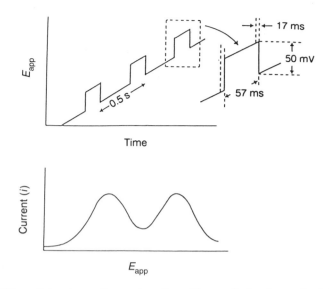

Figure 4. Differential pulse voltammetry. **Top:** The applied voltage (E_{app}) is a linearly increasing ramp upon which small amplitude (50 mV) pulses are superimposed (in this case twice per second). The voltage immediately before the pulse is subtracted from that toward the end to give the differential measurement. **Bottom:** The differential current is plotted against the applied voltage to give peaks for each compound separately oxidized.

duration (30–50 mV). The current is sampled immediately before a pulse and subtracted from the current at the end of the pulse, and this difference expressed in terms of potential. *Differential pulse voltammetry* (DPV) combines the main advantage of chronoamperometry (suppression of charging current) with the resolution of linear sweep voltammetry, in essence performing a local differentiation of the voltammogram obtained by linear sweep voltammetry (*Figure 4*). The overlap between two oxidizable compounds is eliminated, providing that they oxidize at sufficiently distinct potentials (at least 100 mV), resulting in high resolution (6). The sensitivity and resolution can be manipulated by altering the characteristics of the pulses. Increased pulse amplitude increases the sensitivity but decreases peak resolution. Slower scan rates improve both sensitivity and resolution, but can have the disadvantage of long sampling times resulting in depletion of the electroactive substances in the vicinity of the electrode. Thus it is essential to allow adequate time between each sample measurement for diffusion to restore the levels of the electroactive compounds around the electrode to their equilibrium value. DPV provides both qualitative and quantitative information on the electroactive compounds.

A modification of DPV is differential double (or normal) pulse voltammetry (DDPV or DNPV) in which the applied potential waveform is a double potential step, returning to the initial (resting) value between each of the double pulses (*Figure 3*). The advantage of this approach is that less electroactive material is oxidized than with DPV so less reaction product is adsorbed on to the electrode surface thereby giving increased electrode lifetime.

This chapter will focus solely on the methodologies of FCV and DPV as these are the methods most in use at the present time for investigating extracellular catecholamines and indoleamines and their metabolites.

1.3 General problems of using voltammetric electrodes for measurement of extracellular brain amines and metabolites

It is essential to understand that *in vivo* voltammetric electrodes only record concentrations of electroactive compounds within *extracellular fluid*—none in common use measure *intraneuronal* levels in the mammalian brain. The choice of electrode is central to the successful use of any voltammetric technique and certain essential features are required.

1.3.1 Electrode size

Existing electrodes range in diameter from 1–1000 μm approximately, while synapses are 1–10 nm. No electrodes monitor in the synapse, but rather detect changes in the overflow of either transmitter or metabolites into the extracellular space. The electrodes described in this chapter are all at the low end (1–30 μm) of the diameter range quoted above.

1.3.2 Electrode composition

The oxidation process occurs at the surface of the electrode (*Section 1.1*), and experience has shown that carbon is the most practical material for measuring amines and their metabolites. Various forms of carbon have been employed:

Carbon paste was used for the original electrodes; they are easy to make and stable over long time periods but have several disadvantages, in particular their size (up to 500 μm) (7).

Graphite–epoxy (8) produces hard and durable electrodes but with much the same limitations as carbon paste.

Carbon fibres are used in all the electrode types described in this chapter as they show certain distinct advantages over carbon paste or epoxy electrodes.

First, the diameter is reduced (usually 8–30 μm) such that it may be implanted in small brain nuclei like the raphe system (9–11) or the suprachiasmatic nucleus (12, 13). However, the *active surface* at least equals that of the large paste electrodes. Very small electrodes, have good mass transport characteristics and thus high sensitivity (14). In addition, the active surface of the fibre electrode (the protruding tip) is made of pyrolitic graphite (akin to pure graphite crystal) and, therefore, exhibits excellent electrochemical qualities (14).

Secondly, while the active surface of all other electrodes (paste or disc) is *perpendicular* to the axis of the tissue penetration, with carbon-fibre electrodes almost the whole surface is *parallel* to the penetration axis. This prevents the

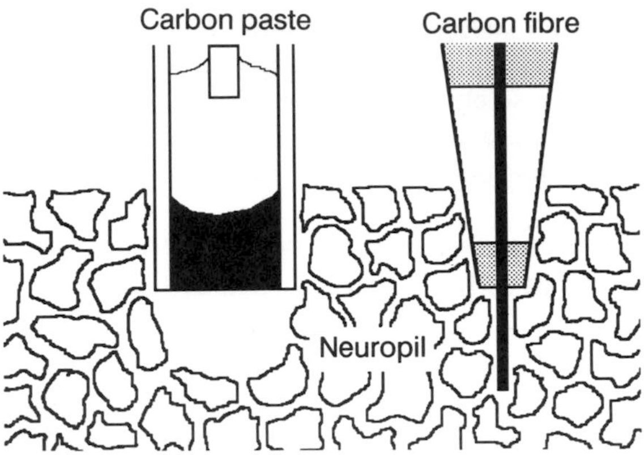

Carbon paste   Carbon fibre

Neuropil

Figure 5. Diagram showing the tips of carbon-paste and carbon-fibre electrodes implanted in the neuropil. A pool of extracellular fluid forms at the tip of the carbon-paste electrode as the tissue is forced apart. The carbon-fibre electrode causes less tissue distortion due to its finer tip.

active surface becoming coated with tissue fragments during *in vivo* implantation thus markedly reducing tissue damage (*Figure 5*). Finally the fast response time of the fibre electrode allows accurate kinetic characterization of the release process measured.

Carbon-fibre electrodes essentially take two forms—cylindrical or disc-shaped. The former were first described by Gonon and co-workers (15–17) and consist of one or more 8 μm o.d. pyrolytic graphite fibres supported in a pulled glass capillary, the active surface of which is a 0.5 mm length of fibre(s) protruding from the capillary. The electrodes are capable of distinguishing between the oxidation of ascorbic acid and catechols after an electrochemical pretreatment (Section 2.1.3). A similar cylindrical, bevelled or conical (but untreated) carbon-fibre electrode is used for FCV (Section 2.2.2).

By cutting the fibre flush with the glass capillary one obtains a very small disc-shaped active surface; such electrodes have been used for *in vivo* studies (7, 18). The extremely small size results in unique electrochemical properties. The catalytic reaction between ascorbic acid and the oxidized form of the catechols, which causes the problem with carbon-paste electrodes, is abolished since extensive spherical diffusion with such small electrodes results in the oxidized product diffusing away from the electrode before it can undergo reaction with ascorbic acid to reform DA (7). Although these electrodes have been used, in combination with normal pulse voltammetry and chrono-amperometry, to monitor extracellular DA resulting from electrically-induced stimulation of DA neurons, (for example ref. 18) they are unable to measure *basal* extracellular levels of DA or indeed DOPAC. Similar electrodes have been used with FCV to monitor electrically-stimulated release of DA in the striatum on a millisecond time scale (19) or for quantification of iono-phoretically applied amine neurotransmitters (20–23). In the latter case the same etched fibre electrodes were used for electrophysiological unit recording.

120

1.3.3 Electrode sensitivity

The importance of sensitivity relates to the interpretation of the voltammetric signal and depends upon what is to be measured with the electrode. Thus an electrode sensitive to micromolar concentrations of DA will not be able to measure *basal extracellular* DA levels, as it is present in nanomolar amounts, but should be able to monitor release following electrical or potassium stimulation (i.e. FCV, Section 2.2). However, an electrode sensitive to low micromolar levels of the dopamine metabolite DOPAC should be able to detect basal extracellular levels of the metabolite as this is present at such concentrations. As a rule of thumb amine transmitters are present at low nanomolar concentrations, so any electrode purporting to measure basal (non-stimulated) transmitter must not only have adequate sensitivity but also be able to detect the amine *without* interference from other electroactive compounds with similar oxidation potentials present in higher (micromolar) concentrations.

1.3.4 Electrode selectivity

There are several compounds in the brain with oxidation potentials within the usable potential window (*Table 1*), so electrode selectivity is central to the development of carbon-fibre electrodes for use with any voltammetric measurement technique. With regard to the catechol and indoleamines and their metabolites the major considerations are:

● Characterization of a catechol peak (DA, noradrenaline (NA), DOPAC) clearly separated from that of ascorbic acid (oxidation potential normally close to that of the catechols, *Table 1*).

● Characterization of an indole peak (5-HT, 5-HIAA) clearly separated from that of ascorbic acid, catechols, and uric acid (oxidation potential similar to indoles, *Table 1*). Carbon-fibre electrodes used for DPV and FCV are able to achieve both the above (Section 2.1.3).

● The second stage in the development of selective electrodes is to achieve separation between catecholamines and their metabolites, and between 5-HT and its metabolite 5-HIAA. There is now available a carbon-fibre electrode that appears to monitor 5-HT without interference from 5-HIAA (24).

● The final stage in selectivity is to develop electrodes specific for particular catecholamines (for example able to monitor DA without interference from NA or DOPAC). This has not yet been achieved.

2. Methodology

2.1 Differential pulse voltammetry (DPV)

DPV combined with electrically pretreated carbon-fibre electrodes provides a reliable method for *in situ* measurement of electroactive compounds. The method provides adequate sensitivity with high resolution between peaks produced by individual compounds. The limitations of the technique mainly relate to the slow scan time (see Section 1.2) and the resulting problem of depletion of electroactive substances in the region of the electrode. This means that adequate time must be left between each scan (60–120 s) for equilibrium to be re-established.

Carbon-fibre electrodes for use with DPV can be used to:

- monitor extracellular ascorbic acid, DOPAC, 5-HIAA (with about 10–30% interference from uric acid) and homovanillic acid (HVA) either simultaneously or individually at the same site (12, 15, 25–29),

- monitor extracellular DA in the striatum or nucleus accumbens following inhibition of monoamine oxidase (MAO) to remove the metabolites (28, 30–32) and

- monitor basal extracellular 5-HT (24, 28).

2.1.1 Equipment for DPV

Successful *in vivo* DPV measurements depend upon two factors:

- Suitable measurement equipment to apply the oxidation potential to the electrode and measure the current resulting from oxidation.

- The type of working electrode used together with suitable auxiliary and reference electrodes.

Measurement equipment

There are various commercial options available for *in vivo* DPV measurements with small carbon-fibre electrodes—i.e. they measure low levels of current with satisfactory sensitivity (nA). Several of the highly sophisticated computer-linked systems available *do not have adequate technical capabilities to achieve satisfactory* in vivo *sensitivity*—check before purchase.

If it is proposed to use electrically pretreated carbon fibres there are two options:

- Purchase a polarograph that does not have a built-in function generator necessary for the electrode pretreatment and obtain these items separately, for example Princeton polarograph (EG & G, UK) plus a waveform generator (for details see section on electrode pretreatment).

- Purchase a Tacussel 'Biopulse' (Roth Industrial Instruments Ltd, UK) which includes the necessary waveform generator.

The polarograph needs to be linked to a suitable recording device—the simplest is a pen recorder with manual measurement of the oxidation peaks. The alternative is to use a microcomputer, but this will require suitable programming.

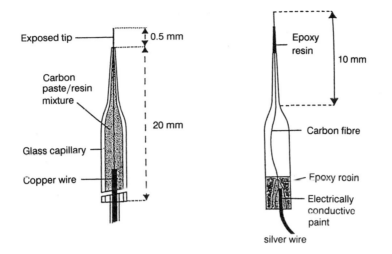

Figure 6. Diagram showing the basic construction of carbon-fibre electrodes for use with differential pulse voltammetry (DPV). **Left:** The design of the original electrode (Ponchon *et al.* 1979) which used carbon paste to make reliable contact between the fibre and the copper wire. Tip length is 0.5 mm, fibre diameter is 8, 12 or 30 µm. **Right:** The modified electrode, described in this chapter, which uses epoxy resin and electrically conductive paint to make electrical contact.

2.1.2 Electrodes for DPV

Working electrode
Details of the manufacture of a basic carbon-fibre electrode suitable for DPV based upon Ponchon *et al.* (17), but extensively modified (29), are given in *Protocol 1* and shown in *Figure 6*.

Protocol 1. Basic carbon-fibre electrode for DPV

1. Pull a length of glass capillary (Clark Electromedical Instruments, 1.2 mm o.d., 0.69 mm i.d., ref. GC150–15) to a fine tip using a conventional electrode puller and cut to a length of approximately 25 mm.

2. Isolate 1–3 carbon fibres, for example AVCO, Lowell, USA—30 µm diam. from a bunch, thread into the pulled capillary and push to the tip.

3. Cut the tip of the capillary using a pair of iris scissors such that the fibre can be pushed through the end to protrude approximately 5 mm. Cut the opposite end of the fibre flush with the capillary.

4. Dip one exposed end of a length of Teflon-coated silver or plastic-coated copper wire in electrically conductive paint (Radio Spares) and place in contact with the carbon fibre.

5. Strengthen this connection with polyester resin.

Protocol 1. *Continued*

6. Seal the tip of the electrode by applying a drop of low viscosity resin (Loctite Glassbond) using a piece of thin wire taking care to avoid fouling the exposed fibre with resin. Capillary action ensures a good seal between glass and fibre. Allow to harden overnight.

7. Cut the exposed fibre to length (0.5 mm or 0.3 mm) immediately prior to use.

In most studies electrodes made of either two or three carbon fibres are used to increase the amplitude of the signals recorded and produce a stronger electrode tip that is less likely to break during implantation. Electrodes with 'splayed-out' fibres should be discarded. This basic carbon-fibre electrode must then be electrically pretreated to produce electrodes that separate ascorbic acid, catechols, and indoles *in vivo* with DPV (see Section 2.1.3).

A development of the basic carbon-fibre electrode is the Nafion-coated carbon-fibre electrode. Nafion is a sulphonated polymer which repels anions but is selectively permeable to cations. Thus the combination of electrical pretreatment and Nafion coating should provide an electrode able to detect the catechol (DA, NA) and indoleamines (5-HT) rather than their metabolites (for example DOPAC, 5-HIAA) providing there is *adequate sensitivity* for the amines (24). The basic construction of Nafion-coated electrodes is given in *Protocol 2*.

Protocol 2. Nafion-coated carbon-fibre electrode

1. Follow 1–7 of *Protocol 1*.

2. Clean the carbon-fibre tip of the electrode by dipping into chromic acid (5 g sodium dichromate in 5 ml water and 95 ml concentrated sulphuric acid) followed by thorough washing in distilled water.

3. Pretreat electrically as described in *Protocol 7*.

4. Immerse the electrode tip into a drop of Nafion (10 µl of a 5% solution: C. G. Processing, Delaware, USA) placed in a 3 mm loop at one end of a platinum wire (approx 100 µm diameter). Connect the other end of the wire to the reference and auxiliary outputs of the polarograph and apply a DC potential (+ 3.7 V) for 2 s. Repeat this 4 times.

5. Wash the electrode in distilled water and oven dry at 60 °C for 20 s.

6. Use immediately for *in vitro* and *in vivo* recordings.

In an attempt to obtain a more selective and sensitive amine-selective electrode, a modified version of the Nafion carbon-fibre electrode is being developed, called the NA–CRO carbon-fibre electrode with an active tip (30 µm in diameter) coated with a 50/50 mixture of Nafion and dibenzo-18-crown-6 (Aldrich, Gillingham, UK).

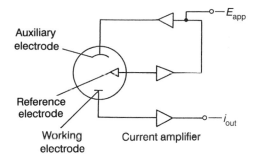

Figure 7. Diagram showing the three electrode potentiostat used for voltammetric experiments. When a voltage (E_{app}) is applied to the potentiostat current flow through the working electrode (i_{out}) consists of charging and faradaic current.

This improves the negative electrical charge of the coat while inserting in it a chemical trap for cations formed by the negative crown structure of the second compound.

Reference and auxiliary electrodes

Electrochemical detection involves the application of a potential to the working electrode with respect to a reference, with concomitant measurement of the current generated by oxidation or reduction. In practice this necessitates the use of at least two, but generally, three electrodes (*Figure 7*). The potential is applied between the oxidizing electrode (the working electrode) and a reference half-cell (normally Ag/AgCl) (*Protocol 3*) against which all applied potentials are expressed. All voltages quoted in this chapter are relative to Ag/AgCl. A third, auxiliary electrode (platinum, silver, or copper) is added to provide potentiostatic control. The potentiostat is normally composed of three operational amplifier sections; a signal portion generates the applied potential, and a control amplifier maintains the potential difference between working and reference electrodes by passing the required current via the auxiliary electrode. An output section measures the current generated as a result of the oxidation or reduction processes. The commercial polarographs listed in Section 2.1.1 and 2.2.1 provide the potentiostatic control. The auxiliary electrode plays no role in the measurement. It is simply a part of the control system preventing current flow between the working and reference electrodes and may be a length of platinum for *in vitro* work or silver, copper, or platinum for *in vivo* experiments.

Protocol 3. Reference electrode

1. Take 30 mm of Teflon-coated silver wire (Clark Electromedical Instruments) and strip 5 mm of the Teflon coat from each end.

2. Connect one end to the positive pole of a regulated power supply. Immerse the other end in 1 M HCl.

Protocol 3. *Continued*

3. Immerse one end of a length of copper or platinum wire into the acid and connect the other to the negative pole. This completes the circuit.

4. Apply $+5$ V for 1 min—this will coat the exposed silver wire with a thin layer of silver chloride.

2.1.3 Electrode pretreatments for DPV

Since *untreated* carbon-fibre electrodes are not suitable for DPV, various electrical pretreatments are used to improve electrode selectivity and sensitivity. The first enables the separation of the oxidation of ascorbic acid from catechols (16) (*Protocol 4, Figure 8c,d*).

Protocol 4. Electrical pretreatment to separate ascorbic acid and catechol oxidation potentials

1. Immerse the working, reference, and auxiliary electrodes in phosphate-buffered saline pH 7.4, 0.1 M and connect to the relevant terminals of the polarograph.

   ● If the polarograph is a Tacussel 'Biopulse', the signal generator is an integral part of the instrument.

   ● If the polarograph is a *Princeton 174A type* the output and earth terminals of a signal generator (GX 115 Metric) must be connected to the polarograph via pins 9 and 10 of plug J38. Apply a 70 Hz triangular wave potential (0 to $+3.4$ V), for 20 s with the polarograph in the 'Sampled DC' mode, initial potential zero. Disconnect the signal generator.

2. With the polarograph in the 'DC' mode apply the following potentials sequentially:

   ● $+1.5$ V for 5 s

   ● -0.9 V for 5 s

   ● $+1.5$ V for 5 s[a]

3. Wash electrode in distilled water and oven dry (60 °C) for 1 h.

[a] Note that it may be necessary to vary the last potential between 1.1–1.5 V depending on the sensitivity of the electrode.

The optimum conditions for the separation of an indole (+ uric acid) peak from other oxidation products involve a slightly modified version (9, 34) of *Protocol 4* (*Protocol 5, Figure 8a,b*).

Protocol 5. Electrical pretreatment to separate the oxidation of indoles from catechols and ascorbic acid

1. Follow step 1 in *Protocol 4*, but modify the triangular waves applied to the following:

   - 0 to +3 V at 70 Hz for 20 s
   - 0 to +2 V at 70 Hz for 20 s
   - 0 to 1 V at 70 Hz for 20 s

2. Follow step 3 of *Protocol 4*.

Protocol 5 will not produce separation of ascorbic acid and catechols which form a merged peak but will give a clear separation from the indole peak (Section 2.1.4).

More recently, the pretreatment has been further modified to produce an electrode that can monitor ascorbic acid (peak at -50 mV) catechols ($+80$ to 100 mV) and indoles ($+300$ mV) simultaneously as three distinct peaks (26, 35) as well as a fourth peak ($+400$ to 420 mV) (36) (*Protocol 6, Figure 8e,f,g,h*).

Protocol 6. Electrical pretreatment to separate ascorbic acid, catechol and indole oxidation peaks

1. Follow step 1 in *Protocol 4*, but modify the triangular waves applied to the following:

   - 0 to 3 V at 70 Hz for 10 s
   - 0 to +2.5 V at 70 Hz for 15 s
   - 0 to +1.5 V at 70 Hz for 20 s

2(a). If using the Princeton 174A disconnect the signal generator and switch to 'DC' mode. Apply the following continuous potentials.

   - +1.5 V for 5 s
   - -0.9 V for 5 s
   - +1.5 V for 5 s

2(b). If the Tacussel 'Biopulse' polarograph is used follow the procedure below:

   - triangular wave at 70 Hz from 0 to +2.8 V for 8 s
   - two successive continuous potentials applied +1.35 V for 5 s; -0.75 V for 5 s

3. Follow step 3 of *Protocol 4*.

It is still unclear why the electrical pretreatment increases the sensitivity and selectivity of the carbon fibres. The effects on the chemical and structural features of the fibres have been studied using transmission and scanning electron microscopy. Results show that the pretreatments may increase the redox couples and oxides both on the surface and within the core of the fibres so enhancing their interaction with electroactive compounds. As yet there is no rationale that can be applied to produce the correct pretreatment parameters for a particular electroactive compound—serendipity is still the way ahead.

Nafion-coated electrodes require a different pretreatment regime. With reference to *Protocol 2*, step 3, these electrodes should be electrically pretreated using *Protocol 7* to obtain maximum selectivity and sensitivity to 5-HT (*see Figure 9*).

Protocol 7. Electrical pretreatment of the Nafion-coated carbon-fibre electrode

1. Follow *Protocol 2* to step 3

2. Apply the following 70 Hz triangular waves in order:
   - 0 to +3 V for 15 s
   - 0 to +2.5 V for 20 s
   - 0 to +1.5 V for 30 s

3. Apply three successive continuous potentials:
   - +1.25 V for 10 s
   - -0.75 V for 10 s
   - +1.3 V for 10 s

4. Wash the electrode in distilled water, oven dry (about 60 °C) for 10 min.

2.1.4 DPV parameters

The normal DPV scan parameters for both *in vitro* and *in vivo* measurements with the Princeton 174A are:

- potential range: -0.2 V to +0.45 V
- scan rate: 5 mV/s
- pulse amplitude: 50 mV
- pulse frequency: 2 or 2.5 per s
- pen filter set at 0.3–1

Follow the manufacturer's instructions for setting up these parameters. Remember that the potential range will determine how many peaks you detect and should be *modified* depending on need. For example, if only ascorbic acid and catechols are being monitored do not go above +0.35 V.

Table 2. Recommended concentrations of electroactive substances for *in vitro* testing of DPV carbon-fibre electrodes

Substance	Concentration (M)
Ascorbic acid	10^{-5}
Dopamine	10^{-8}
DOPAC	10^{-6}
5-HT	10^{-8}
5-HIAA	10^{-7}
Uric acid	10^{-6}
HVA	10^{-7}

Individual solutions and appropriate mixtures should be prepared in artificial CSF or phosphate-buffered saline pH 7.4.

It is necessary to leave adequate time between each scan to allow the extracellular concentrations to re-equilibriate by diffusion. The length of time depends, in part, on the type of study but normally 2 min is adequate. Between scans the potential should be held at the initial potential for the scan (i.e. -0.2 V).

2.1.5 Testing and calibration of DPV electrodes

Testing
After preparation of the electrodes the next step is to determine whether the electrical pretreatment has produced the expected separation of the different oxidation peaks. The simplest way is to test the electrodes in standard solutions of ascorbic acid, DA, DOPAC, 5-HT, 5-HIAA, uric acid, and HVA made up in artificial CSF (pH 7.4) at concentrations relevant to their expected levels in extracellular fluid (*Table 2*). Electrodes not showing their expected properties should either be electrically pretreated again or discarded. Expected separation of the oxidation peaks using the electrical pretreatments given in *Protocols 4–6* are shown in *Figures 8 and 9*.

A good carbon-fibre electrode for DPV measurements has low resistance (300–800 Ω), so a low 'charging current' is obtained when the electrode is switched on. Very low noise and a stable background current (thus a stable baseline between measurements) are also characteristics of a good electrode.

Calibration
Calibration of DPV carbon-fibre electrodes so that comparison between *in vitro* sensitivity and *in vivo* extracellular levels can be made is a major difficulty. Recalibration *in vitro* after making *in vivo* measurements demonstrates that the *in vitro* sensitivity of the electrode may alter following a period *in vivo* and thus calibration *in vitro* **before** and **after** use *in vivo* are essential.

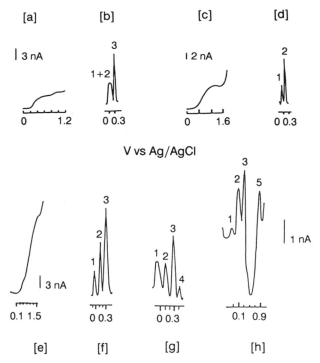

Figure 8. Typical DPV scans showing the effects of the various pretreatments upon electrode sensitivity and selectivity. [a] and [b] *In vitro* voltammograms recorded in a mixture of ascorbic acid (0.5 mM), DOPAC (50 μM) and 5-HIAA (25 μM) at an electrode before [a] and after [b] pretreatment using the schedule described in *Protocol 5*. Note that the 5-HIAA peak (3) is clearly separated from the composite ascorbate and DOPAC peak (1 + 2). No resolution is observed prior to treatment. [c] and [d] As in [a] and [b] but with the electrode exposed to the treatment schedule described in *Protocol 4*. Ascorbate (1) and DOPAC (2) are clearly separated. 5-HIAA was not measured in this example. [e] and [f] As in [a] and [b] but with the electrode pretreated using *Protocol 6*. Note the separation of ascorbate (1), DOPAC (2), and 5-HIAA (3) peaks. [g] and [h] When electrodes pretreated according to *Protocol 6* are implanted in the rat striatum up to 5 peaks may be detected according to the range of scan: ascorbate (1), DOPAC (2), 5-HIAA (3) homovanillic acid (4), and tryptophan/tyrosine/peptide (5). The best recordings are obtained by stopping the scan at about 0.45 V so that the final peak is not recorded.

Thus it is difficult to calibrate accurately the carbon-fibre electrodes for *in vivo* use, but it is essential to check *in vitro* the separation characteristics and relative sensitivities towards the compounds to be detected *in vivo*. The carbon-fibre electrodes provide reliable **qualitative** and interesting **semi-quantitative** information.

Electrode life
Carbon-fibre electrodes show a useful life *in vivo* as long as the separation between the amine metabolites and ascorbic acid achieved by the electrical

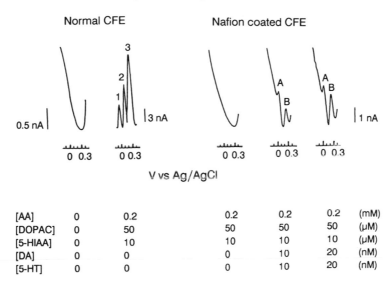

Normal CFE   Nafion coated CFE

V vs Ag/AgCl

[AA]	0	0.2	0.2	0.2	0.2	(mM)
[DOPAC]	0	50	50	50	50	(µM)
[5-HIAA]	0	10	10	10	10	(µM)
[DA]	0	0	0	10	20	(nM)
[5-HT]	0	0	0	10	20	(nM)

Figure 9. An *in vitro* comparison between a normal 12 µm carbon-fibre electrode (CFE) pretreated according to *Protocol 6* and a Nafion-coated electrode pretreated using *Protocol 7*. Signals were recorded in phosphate buffer alone or containing ascorbate (0.2 mM), DOPAC (50 µM) and 5-HIAA (10 µM), or these acids plus DA and 5-HT (10 and 20 nM). Note that the normal CFE exhibits peaks for ascorbate, DOPAC and 5-HIAA while the Nafion-coated CFE does not. The Nafion-coated CFE responds, however, to subsequent addition of the cations DA (A) and 5-HT (B) in a concentration-dependent fashion.

pretreatment is retained. For ascorbic acid and DOPAC measurement the limiting factor is the loss of separation between DOPAC and 5-HIAA peaks after 5–8 hours, so for chronic experiments rather cumbersome replaceable electrode systems are required (34, 37). 5-HIAA electrodes are more stable, the separation lasting up to weeks (10, 11), though two days is a more realistic expectation.

2.2 Fast cyclic voltammetry (FCV)

The history of fast cyclic voltammetry (FCV) is rather different from that of other *in vivo* voltammetric techniques. Whereas the remainder are essentially standard electroanalytical methods that might be found in a typical analytical chemistry laboratory, FCV was created in a neurophysiological environment.

In 1979 Millar and Armstrong James created a novel electrophysiological microelectrode from single carbon fibres encased in tapered glass capillaries (20). The resulting electrode had an excellent signal to noise ratio (comparable with tungsten) and, because of its long taper, caused less damage to the tissue (see *Figure 5*). At the same time, in France, Ponchon and co-workers reported the initial use of carbon fibre for electrochemical studies (17). They described an electrode that could detect catecholamines *in vitro* with the prospect that this might be applied *in vivo*.

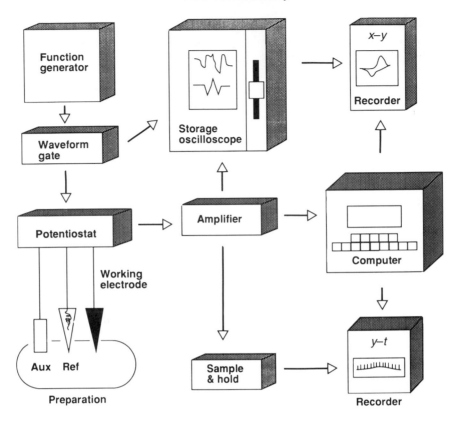

Figure 10. FCV instrumentation (schematic). The function generator produces a continuous triangular output which is gated into discrete 1½ cycle waveforms and fed to the potentiostat. Current output from the working electrode is amplified and displayed on a digital storage oscilloscope. A sample and hold circuit is set to monitor current at a defined potential and its output is displayed on a *y–t* chart recorder. The *x–y* recorder displays cyclic voltammograms. A microcomputer may both receive and process data prior to display.

Although initial work *in vitro* used linear sweep voltammetry, it was quickly recognized that reversing the scan after oxidation (cyclic voltammetry) yielded reduction peaks that characterized the compounds further. The second major change was to increase the scan rate dramatically. Linear sweep scans took several tens of seconds and, from a neurophysiological standpoint, this was too slow. Thus the scan rate was increased from 10 mV/s to 300 V/s. The technique was originally christened high speed cyclic voltammetry (HSCV) although the term FCV is now accepted (38).

The assets of the method are excellent spatial and temporal resolution, as well as having a reduction scan which aids identification of compounds and helps maintain an electrochemically unperturbed environment *in vivo*. FCV is mainly used to monitor rapid stimulated monoamine release and uptake events *in vivo*.

2.2.1 FCV equipment

FCV can be approached from two angles: 'off-the-peg' or 'do-it-yourself'. If one has a good electronics workshop it is possible to assemble the equipment from components. The FCV circuit diagrams (39) enable one to produce a basic set-up and this represents the cheapest route. *Figure 10* shows a schema of the apparatus. A function generator produces a continuous triangular waveform. This is gated into 1½ cycle segments at predetermined intervals (see Section 2.2.4 Voltammetric parameters for FCV) and applied to the potentiostat. Output from the working electrode is then amplified as appropriate and, together with the input waveform, displayed on a digital storage oscilloscope. The oscilloscope should have the facility to subtract waveforms from one another. A sample and hold circuit gives a continuous record of oxidation (or reduction) current at a single potential and acts as a readout of the concentration of the compound during the experiment. The output from the 'sample and hold' is displayed on a standard y–t chart recorder. These two aspects of the method are complementary: the oscilloscope allows processing of entire waveforms (voltammograms) to compare with standards, thus providing the 'fingerprint' of the compound while the 'sample and hold' provides information about concentration dynamics (see also *Figure 12*).

The simpler approach is to buy one of two polarographs designed specifically for FCV: The Millar Voltammeter (PD Systems, UK) or Ensman EI 400 (Ensman Instrumentation, USA). Each incorporates the function generator, potentiostat, sample and hold, amplifiers and filters. All one then needs is an oscilloscope and a chart recorder.

2.2.2 Electrodes for FCV

The only working electrodes used *in vivo* with FCV are carbon fibres. However, there is great diversity in the types of fibre and in their preparation and pretreatment.

Carbon fibre is available from many sources. Each company has its own recipe and the resulting fibres differ in carbon content, diameter, and surface chemistry. This is worth bearing in mind when comparing results from different laboratories using otherwise similar protocols. Wightman's group in the USA has reported differences in the electrical properties of some of the most commonly used fibres (7).

Carbon fibre is available in various diameters, from 4 to 40 μm, although our experience has been gleaned using 8 μm diameter fibres (Courtaulds XA-S). A protocol detailing the preparation of carbon-fibre microelectrodes for unit activity recording and FCV is to be found in Chapter 1 (*Protocol 3*).

Although our standard type of carbon-fibre microelectrode for FCV has a cylindrical tip, it is possible to prepare other types. One can, for example, bevel the surface with diamond paste or etch the tip in chromic acid. *Figure 11* shows various types of electrode tip configuration.

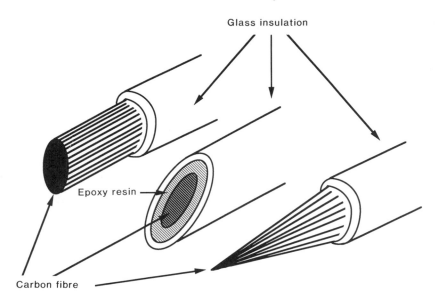

Figure 11. Carbon-fibre microelectrodes for FCV. Three different types of carbon-fibre microelectrode tips. **Left:** Cylindrical microelectrode in which the carbon fibre protruding beyond the glass insulation is trimmed, under micromanipulator control, to the desired length (typically 30–50 μm). **Centre:** Bevelled tip produced by placing a pulled microelectrode at a shallow angle on to a revolving plate covered in diamond paste. Excess paste is removed by rinsing in toluene. **Right:** Conical electrode tip produced by electrically etching the protruding fibre in dilute chromic acid.

As well as the working electrode you will also need an auxiliary electrode which applies the voltage. As stated earlier (Section 2.1.2), this takes no direct part in reactions at the working electrode and need be nothing more sophisticated than a copper, silver, or platinum wire.

The reference electrode is more important. It provides a constant point with which voltages at the working electrode are compared. Saturated calomel (involving mercury) is less desirable than the almost universal silver–silver chloride (Ag/AgCl) electrode. These can be bought as small discs (Clark Electromedical) which are then inserted into a plastic pipette tip filled with 0.9% saline as a salt bridge. *Protocol 3* describes how to make your own.

2.2.3 Working electrode pretreatments for FCV

Unlike DPV, where the electrodes are unusable without pretreatment, there is little place for surface treatments with FCV since it is intended to measure very short duration events. Pretreatments usually increase sensitivity while slowing the response time of the electrodes due to adsorption of substrate. Over the time course of DPV measurements (tens of seconds) this is unimportant and the improved sensitivity is a boon. However, with FCV the slowed time response can be a major handicap.

Table 3. Typical FCV operational parameters

	'BASIC'	'EXTENDED'	'RMW'
Negative scan limit	− 1.0 V	− 1.0 V	− 0.4 V
Positive scan limit	+ 1.0 V	+ 1.4 V	+ 0.8/1.0 V
Scan rate	300 V/s	300–500 V/s	300 V/s
Resting potential	− 0.2 V (or disconnected)	− 0.2 V (or disconnected)	− 0.4 V
Number of cycles	1.5	1.5	1
Scan repetition rate	up to 40/s	usually 1–4/s	up to 60/s

In general, brief (30 s) anodic offset of the input waveform (0 to + 2 V) enhances FCV sensitivity by promoting adsorption. The same effect can be achieved by etching the tips in acid dichromate (*Figure 11*). However, in both cases, the electrodes poison badly on contact with brain tissue and the improvements are transitory.

When improved sensitivity is the overriding consideration, a modified input waveform may be used (see Section 2.2.4). This ongoing treatment allows improved sensitivity at some expense to the temporal response. Unlike *pre*treatments, the effects are preserved throughout the experiment.

2.2.4 Voltammetric parameters of FCV

The definable parameters of the FCV input waveform are the positive and negative limits of the waveform, the scan rate, repetition rate, and resting potential between scans. Although each parameter may be altered according to the specific dictates of the experiment, we have found it best to stick to a few well-characterized basic waveforms (see *Table 3*). It should be stressed that these are not the only usable input waveforms. However, when using new waveforms, it is advisable to characterize the sensitivity and time response of the new parameter set thoroughly before experimentation.

The 'basic' waveform (1½ cycles of a 75 Hz triangle from − 1.0 to + 1.0 V) has been extensively used to measure stimulated DA release. The initial negative half cycle tests for the presence of oxidized species at the electrode surface and is followed by the main oxidation (− 1.0 to + 1.0 V) and reduction (+ 1.0 to − 1.0 V) scans. The voltage is then restored to zero or the resting potential. These parameters offer good sensitivity and high temporal resolution, unimpaired by adsorption. The lower limit of detection for DA is about 100 nM. The waveform can be repeated up to 40 times per second without loss of sensitivity. Typically the working electrode is electronically disconnected between scans, although holding at − 0.2 V between waveforms gives identical results.

The 'extended' waveform was developed in recognition of the need for higher sensitivity in certain situations. The difference is that the anodic scan extends to + 1.4 V rather than + 1.0 V. This acts as a continuous treatment and improves

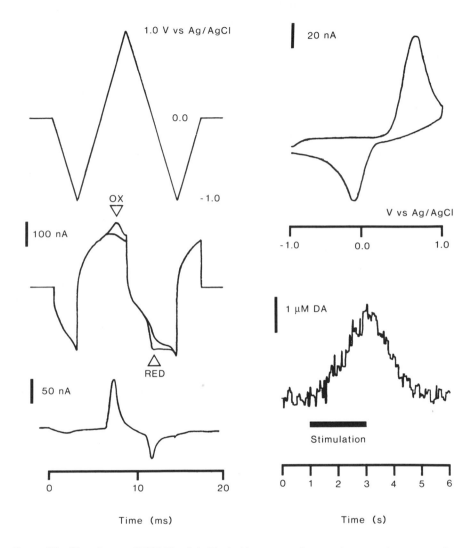

Figure 12. Waveforms of FCV. **Top left:** Typical input waveform to the potentiostat scanning from − 1.0 to + 1.0 V vs Ag/AgCl (300 V/s). **Middle left:** Two superimposed current output waveforms obtained in buffer before and after the addition of DA (20 μM). Note that the signals are exactly superimposable except at the DA oxidation and reduction peaks (△). **Bottom left:** Faradaic current for DA oxidation and reduction obtained by electronically subtracting the charging current (signal in buffer) from the test signal. **Top right:** Faradaic current for DA oxidation and reduction plotted against the input voltage to the potentiostat to form a cyclic voltammogram. **Bottom right:** Sample and hold output at the DA oxidation peak potential from an electrode implanted in the rat striatum during stimulation of the median forebrain bundle (horizontal bar). During stimulation the DA concentration in the extracellular fluid rises as DA is released. On cessation, the DA level falls due to uptake.

sensitivity to DA by almost an order of magnitude (40) although the electrodes become more adsorptive and their time response to changes in DA concentration is, to some extent, slowed. This may make these parameters unsuitable for studies of monoamine uptake.

The 'RMW' waveform is that used in Wightman's laboratory in the USA. It is, in essence, a simplified version of the basic waveform but omitting the initial negative half cycle of the scan. The sensitivity and time response of this parameter set is essentially that of the basic waveform.

At the scan rates used (300–500 V/s) the waveform lasts 10–20 ms and can be repeated at intervals as short as 5 ms. This gives near 'real time' measurement of concentration dynamics (see Section 4.2. Applications of FCV).

When the FCV input voltage is applied to the potentiostat, the working electrode output exhibits very high charging current (*Figure 12*). This occurs irrespective of any oxidizable or reducible compounds being present and is thus a 'background' signal, due to the capacitative nature of the electrode–solution interface. Although large, often dwarfing oxidation and reduction peaks, it is constant and thus can be removed from the electrochemical signals. Typically a 'background' signal is stored on one channel of the oscilloscope. DA, for instance, is then added to the solution (or its release is evoked *in vivo*) and a second 'test' signal is stored on the other channel. The background is then electronically subtracted from the test signal leaving only the faradaic (redox) current from the DA oxidation and reduction. With high resolution oscilloscopes tiny faradaic signals can be readily extracted from charging current signals even a hundred-fold larger (see *Figure 12*). It is worth noting that one must always use an appropriate background signal for subtraction. A background recorded in buffer cannot be used for comparison with *in vivo* signals, for instance.

At the high voltage scan rates used with FCV, compounds typically oxidize at much higher potentials than with DPV. DA, for example, has an oxidation peak at +500 to +600 mV vs Ag/AgCl with FCV compared with +150 to 250 mV using DPV. One should be aware that peak potentials obtained using different methodologies are not directly comparable. The usable voltage window is also expanded with FCV.

2.2.5 Testing and calibration of FCV electrodes

One quickly gets a feel for what makes a good or bad electrode. A good electrode gives sharp peaks for catecholamines, has a rapid response time, stable background current, and low noise. Often one can tell from the background current whether the electrode is viable or not. *Figure 13* shows examples of good and bad electrode signals.

Electrically, an electrode is somewhere between a resistor and a capacitor in behaviour. The good electrode behaves in a more capacitative fashion with sharp transients at the scan reversal points and a rather 'squared-off' shape. Conversely, the bad electrode has a much more resistive shape. This, in our

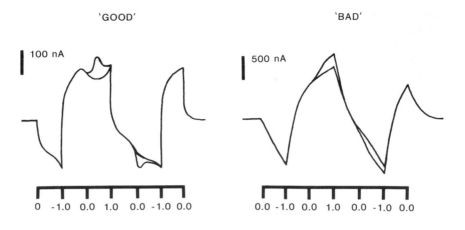

'GOOD' 'BAD'

Input voltage (V vs Ag/AgCl)

Figure 13. Good and bad FCV electrodes. Background current signals obtained in buffer before and after the addition of DA (20 µM). **Left:** A good electrode showing pseudocapacitative signal with clear peaks for DA. **Right:** A bad electrode showing resistive 'spiky' background current and poor definition of oxidation peaks.

experience, is usually due to a poor glass–carbon seal, often leading to a large unstable background current. With practice one learns quickly to reject the poor electrodes.

With any voltammetric experiment one must calibrate the working electrode. With FCV this is usually done *after* the experiment rather than beforehand. Electrode characteristics are often altered by contact with the brain as with the DPV electrodes, the main consequences of this 'poisoning' being a drop in sensitivity, decreased resolution between adjacent peaks, and a shift in oxidation peaks to higher potentials. For FCV these changes are 'instantaneous', occurring during electrode implantation as a result of cell rupture or other related trauma. There is usually no further deterioration: electrodes left *in vivo* for 8 hours poison no more than those removed after 30 minutes. FCV itself does not cause poisoning: electrodes deteriorate to the same degree whether used for FCV *in vivo* or simply implanted without any electrochemical recordings being made. Thus it makes sense to calibrate electrodes after the experiment, since the characteristics of the electrodes are then directly comparable with those *in vivo*. Precalibration should only be used when unavoidable and the results must be interpreted with caution.

2.2.6 FCV and electrical stimulation

FCV is used primarily to measure stimulated monoamine release *in vivo* and *in vitro* (see Section 4.2 Applications of FCV). In each case electrical stimulation is applied via an isolated stimulator to avoid interference with the ground

138

of the voltammetric system. With isolation, electrical crosstalk between systems is negligible. Stimulation currents of 100 µA can be applied without affecting voltammetric signals as much as a thousand-fold smaller.

In vivo, stimulation is usually applied to the axon bundles by bipolar electrodes. Most experience has been gained in the nigrostriatal system where the stimulating electrode is located in the median forebrain bundle. Sinusoidal stimulation, although unfashionable, is highly effective at evoking DA release in the forebrain. Frequencies of 50–60 Hz are optimal at RMS current values of around 100 µA. Similar results are obtained using biphasic pulses (300 µA, 60 Hz, 2 ms duration).

In vitro (brain slices) stimulation is applied more locally, about 200 µm from the working electrode. Most work has used pulses (0.1 ms) since sine wave stimulation causes more interference. In striatal slices DA release can be evoked even from single pulses of stimulation although in other nuclei, with other transmitters, trains are necessary.

In each new situation it is necessary to assess stimulation parameters empirically. There is no reason to believe that any other than the most general rules apply for other nuclei or neurotransmitters.

2.3 Preparations used with voltammetry

2.3.1 Brain slices

Voltammetry in brain slices differs from classical superfusion studies of transmitter release from slices in that the preparation cannot be enclosed by the chamber. It is necessary to leave access for the voltammetric electrodes. We have found the open top bath design (41) to be ideal. In this design a 350 µm slice is supported on a platinum grid (which also serves as the auxiliary electrode) and held in place by a nylon mesh. The method has been described in detail elsewhere (42) but a few points are worth mentioning.

- Insert the voltammetric electrode tips to a depth of about 80 µm below the surface of the slice. This ensures that they do not record from damaged tissue yet still receive sufficient oxygenation.

- The temperature found to be best is 32 °C. Higher temperatures lead to less efflux (probably due to greater amine uptake) and a shorter lifetime for the tissue.

- Try to align the stimulating electrode such that it straddles the fibre bundles (where visible). Place the working electrode in the centre of the bundle about 200 µm 'downstream' of the stimulating electrode.

- The electrodes are very sensitive to movement. A solid platform or floating table helps eliminate shocks.

2.3.2 Voltammetric measurements in anaesthetized animals

Choice of anaesthetic
Several DPV studies indicate that the basal levels of the DOPAC and 5-HIAA signals (Peaks 2 & 3) *in vivo* can be affected both by the type of anaesthetic used and the level of anaesthesia. We have studied the effects of both of these, as well as the responses to various drugs of the striatal DOPAC oxidation peak in rats under various anaesthetics (halothane, α-chloralose, pentobarbitone, and chloral hydrate) (43).

For routine use halothane (2–3% in 50/50 O_2/N_2O (1 litre/h)) produces stable baseline DOPAC levels upon which significant changes induced by DA agonists and antagonists are clearly seen. In practical terms, the use of halothane allows rapid induction and constant anaesthesia (see also Chapter 6). The steady DOPAC baseline may reflect the well-balanced nature of the anaesthesia induced by halothane; this contrasts with the situation observed with pentobarbitone. While α-chloralose produces a stable baseline and large drug-induced responses, the physical characteristics of the anaesthesia induced (i.e. certain reflex irritability retained) coupled with the inconvenient administration of the drug, bring into question its suitability as a general anaesthetic for rats. The mechanisms involved in the effects of the different anaesthetics warrant further investigation but barbiturates should be avoided in experiments monitoring amine neurotransmission.

Experimental procedure
Normal stereotaxic procedures are observed using an appropriate rat atlas to obtain regional co-ordinates. A hole is carefully drilled through the skull and a clean hole made through the dura. The working electrode is then carefully lowered through the hole, *taking care not to damage the fibre tip*, to the appropriate level in the brain. For FCV, where a stimulating electrode is also implanted, it is important to position the working electrode *before* the stimulating electrode. The auxiliary and reference electrodes are placed on the dura surface of the parietal area of the cortical bone through a hole in the bone. All the electrodes may then be fixed in place with dental cement and connected to the polarograph.

It is important to maintain anaesthesia at a constant level throughout the experiment (which can last for 6–10 h) and voltammograms should be recorded at regular intervals (for example 3 min for DPV, 1 s for FCV) thoughout. An initial period of at least one hour is required for the signals to stabilize; tissue damage, caused by implantation, results in abnormally high extracellular levels of catechols and indoles. This should be followed by an appropriate period of baseline measurements prior to any drug intervention. For DPV it is important also to check constantly that the working electrode continues to show good separation between the oxidation peaks. Also, any sudden change in peak height should be treated with caution (altered level of anaesthesia, faulty electrical contacts, electrode movement, 'noise' in the system).

2.3.3 DPV measurements in freely moving animals

For studies in freely moving rats, the three electrodes are fixed to the bone with dental cement and protected with a light plastic cover. Connection to the polarograph via a simple three-way swivel allows electrical contact to be retained with free movement of the rat. The rat is placed in a plastic cage with water and food *ad libitum*. DPV measurements are started at the end of the surgery (about 15–30 min), and the voltammetric scans performed automatically every 2 or more minutes once the signal(s) have stabilized after recovery from anaesthesia. Some electrochemists claim that it is possible to repeat the electrical pretreatment of the electrode *in vivo* to prolong the time of the experiments, although it is not known whether such *in vivo* pretreatment causes local brain damage where the electrode is positioned. An alternative approach is to develop a replaceable working electrode system (34), though the change in electrode with varying sensitivities may complicate interpretation of the data.

At the end of the voltammetric study (anaesthetized or freely moving), the position of the electrode tip in the brain should be marked by making an electrolytic lesion (5 mA, 10 s) using the working electrode. The brain is then rapidly removed and standard histological procedures (staining with cresyl violet dye) performed. The current values of the peaks are determined by constructing a tangent to the shoulder of the peak and measuring the perpendicular height between the tangent and the apex of the peak.

3. Identification of voltammetric signals

Voltammetry differs from dialysis in that compounds are measured *in situ*. No samples are removed from the milieu to allow direct verification of identity. Thus it is necessary to use a series of tests to be confident of the identity of the measured species. This approach has evolved from a decade's experience in which some of the early claims of voltammetry subsequently had to be modified in the light of further experiments. At a recent workshop of the leading *in vivo* electrochemists (44), seven operational criteria were advanced for the identification of signals (see *Table 4*).

Clearly, not all of these criteria are relevant to *every* experimental situation. Substrate-specific enzymes do not exist for all compounds. Sometimes a compound is not associated with a clearly defined afferent pathway or its detailed pharmacology is unknown (for example ascorbic acid), and thus this yardstick cannot be used. Nevertheless, one should seek to fulfil as many of the criteria as possible.

4. Applications

4.1 Applications of DPV (*Table 5*)

The major application for DPV has been to measure endogenous levels of the DA and 5-HT metabolites (DOPAC and 5-HIAA) and to investigate the effect

Table 4. Identification of voltammetric signals

I. SENSITIVITY	The sensitivity of the electrode should be adequate to measure the concentrations expected *in vivo*.

Example: DOPAC levels in striatum are about 20 µM. Can your electrode detect DOPAC *in vitro* at concentrations of 5–50 µM?

II. SELECTIVITY	The electrode should be able to measure the compound of interest in the presence of other compounds likely to occur *in vivo*.

Example: 5-HIAA and uric acid oxidize at similar potentials. Can your electrode detect 5-HIAA or urate without interference from the other?

III. ELECTROCHEMICAL IDENTITY	The signal detected *in vivo* should be the same as that obtained for its purported constituent *in vitro*. i.e. oxidation and/or reduction peaks should occur at the same potentials.

Example: DA has oxidation and reduction peaks at +600 and −200 mV vs Ag/AgCl with FCV *in vitro*. Does your 'DA' voltammogram detected *in vivo* have peaks at the same potentials?

IV. ANATOMICAL SPECIFICITY	Lesioning of the appropriate afferent pathway should decrease the signal. Stimulation should elevate the peak.

Example: Is your 'DA' peak in the striatum abolished by 6-hydroxy dopamine injection into the substantia nigra? Does nigrostriatal stimulation increase the signal?

V. ENZYMES	Substrate-specific enzymes, locally applied, should decrease the peak height.

Example: Ascorbate oxidase converts ascorbate to the non-electroactive dehydroascorbate. Does microinjection of ascorbate oxidase next to the electrode abolish your 'ascorbate' peak?

VI. PHARMACOLOGY	The signal should respond predictably to several drugs known to affect its levels. Single drug tests are meaningless.

Example: Is your 'DA' peak increased by haloperidol, nomifensine, and amphetamine? Is it decreased by apomorphine and reserpine?

VII. INDEPENDENT VERIFICATION	Where possible, the levels of a compound determined electrochemically should agree with those measured by other methods, for example dialysis.

Example: Dialysis finds extracellular DA levels in striatum around 20 nM. Does your 'DA' peak correspond to a similar level when calibrated?

of drugs *in vivo*. Such data provides information about drug-induced changes in *metabolism*. However, these changes are *not necessarily* correlated with *release* (28). DPV has been used to study changes in DOPAC response to typical and atypical neuroleptics (45, 46), and neurotensin (47) and 5-HIAA in response to autoreceptor agonists and antagonists (13, 28). Electrodes that simultaneously measure ascorbic acid, DOPAC, 5-HIAA, and HVA have also been used to

Table 5. Applications of DPV using carbon-fibre electrodes

Use	Limitations
1 *In vitro* measurement of stimulated transmitter release and metabolism in brain slices	Slow response time compared with FCV
2 Measurement of amine metabolism *in vivo* (for example drug effects)	Excellent information, but need to remember changes in metabolism do not necessarily reflect changes in release
3 Measurement of stimulated dopamine release *in vivo*	Need to remove DOPAC by pretreatment with MAO inhibitor. Nafion-Crown-electrodes may overcome this problem. Slow response time relative to FCV
4 Measurement of basal extracellular 5-HT *in vivo*	Due to low extracellular levels the 5-HT peak is small so easier to measure an increase than a decrease
5 Measurement of basal extracellular ascorbic acid	Functional role of ascorbic acid in brain unknown
6 Measurement of amine metabolism and (release) in the freely moving animal	Electrode life limited as separation between the peaks lost after several hours

investigate neurochemical differences between inbred strains of mice (27, 36). The small diameter carbon-fibre electrodes have an added advantage in that they can be used to monitor changes in small brain nuclei (for example suprachiasmatic nucleus, 12, 13; dorsal raphe nucleus, 24).

The advent of electrodes that can monitor either stimulated DA release (27, 36) or basal extracellular 5-HT (24, 28) offer new opportunities for carbon-fibre electrodes used with DPV.

Finally, DPV in unanaesthetized animals has been used to monitor changes in 5-HT metabolism in relation to circadian rhythms (10, 11, 34). Further improvement in the electrodes should expand their use for the study of the relationships beween neurotransmitter release and behaviour.

4.2 Applications of FCV

The first reported application of FCV was in conjunction with unit activity recording (see also Chapter 1). A long-standing problem of ionophoretic application of amines *in vivo* was that one had little idea what concentration resulted in the extracellular space. By using FCV at the carbon fibre one could simultaneously monitor the concentration of ejected amine and its effect on the cells around the electrode: quantitative ionophoresis (21).

The logical next step was to move from measuring *exogenous* to *endogenous* amines. FCV cannot yet measure *basal* endogenous amine levels *in vivo*, but it has been extensively applied to the measurement of stimulated amine release in striatal and limbic forebrain. Electrical stimulation of the median forebrain bundle has been shown to release DA in the forebrain that can be studied with sub-second time resolution. On cessation of stimulation it is possible to follow the removal (uptake) of DA from the extracellular space in 'real time' (see *Figure 12*). Several studies have examined the factors controlling DA release and uptake *in vivo* (48, 49).

Brain slices from various nuclei have also been examined and, as *in vivo*, the release and uptake of DA has been extensively characterized (42). Other transmitters are beginning to attract attention. 5-HT and NA can be measured by FCV, and there is good evidence that, in specific nuclei, the release of these amines too can be monitored. This is an area in which brain slices, with their greater anatomical accessibility would be expected to take the lead.

As an interesting postscript one should draw attention to Wightman's work with adrenal chromaffin tissue. By using FCV, it is possible to measure catecholamine release from individual cultured chromaffin cells following appropriate chemical stimulation (50). There is no doubt that this is a highly promising avenue for future research since this kind of spatial definition is beyond most conventional transmitter release studies.

References

1. Adams, R. N. (1969). *J. Pharm. Sci.*, **58**, 1171.
2. Adams, R. N. and Marsden, C. A. (1982). In *Handbook of psychopharmacology* (ed. L. L. Iversen, S. D. Iversen, and S. H. Snyder), Vol. 15, p. 1. Plenum, NY.
3. Kissinger, P. T., Refshauge, C. J., Dreeling, C. J., and Adams, R. N. (1973). *Anal. Lett.*, **6**, 465.
4. Zetterstrom, T., Sharp, T., Marsden, C. A., and Ungerstedt, U. (1983). *J. Neurochem.*, **41**, 1769.
5. Bond, A. M. (1980) *Modern polarographic methods in analytical chemistry*. Marcel Dekker, NY.
6. Flato, J. B. (1972). *Analyt. Chem.*, **44**, 78.
7. Dayton, M. A., Brown, J. C., Stutts, K. J., and Wightman, R. M. (1980). *Analyt. Chem.*, **52**, 946.
8. Falat, L. and Cheng, H-Y. (1982). *Analyt. Chem.*, **54**, 2108.
9. Crespi, F., Cespuglio, R., and Jouvet, M. (1982). *Brain Res.*, **270**, 45.
10. Crespi, F. and Jouvet, M. (1984). *Brain Res.*, **272**, 263.
11. Crespi, F. and Jouvet, M. (1984). *Brain Res.*, **299**, 113.
12. Marsden, C. A. and Martin, K. F. (1986). *Br. J. Pharmacol.*, **89**, 277.
13. Martin, K. F. and Marsden, C. A. (1986). *Eur. J. Pharmacol.*, **121**, 135.
14. Blaedel, W. J. and Mabbott, G. A. (1978). *Analyt. Chem.*, **50**, 933.
15. Gonon, F., Buda, M., Cespuglio, R., Jouvet, M., and Pujol, J. F. (1980). *Nature*, **286**, 902.
16. Gonon, F., Fombarlet, C. M., Buda, M. J., and Pujol, J. F. (1981). *Analyt. Chem.*, **53**, 1386.
17. Ponchon, J-L., Cespuglio, R., Gonon, F., Jouvet, M., and Pujol, J-F. (1979). *Analyt. Chem.*, **51**, 1483.

18. Ewing, A. G. and Wightman, R. M. (1984). *J. Neurochem.*, **43**, 570.
19. Millar, J., Stamford, J. A., Kruk, Z. L., and Wightman, R. M. (1985). *Eur. J. Pharmacol.*, **109**, 341.
20. Armstrong James, M. and Millar, J. (1979). *J. Neurosci. Meth.*, **1**, 279.
21. Armstrong James, M., Millar, J., and Kruk, Z. L. (1980). *Nature*, **288**, 181.
22. Kruk, Z. L., Armstrong James, M., and Millar, J. (1980). *Life Sci.*, **27**, 2093.
23. Millar, J., Armstrong James, M., and Kruk, Z. L. (1981). *Brain Res.*, **205**, 419.
24. Crespi, F., Martin, K. F., and Marsden, C. A. (1988). *Neuroscience*, **27**, 885.
25. Crespi, F., Sharp, T., Maidment, N. T., and Marsden, C. A. (1983). *Neurosci. Lett.*, **43**, 203.
26. Crespi, F., Sharp, T., Maidment, N. T., and Marsden, C. A. (1984). *Brain Res.*, **322**, 135.
27. Crespi, F., Martin, K. F., Heal, D. J., Marsden, C. A., Buckett, W. R., and Sanghera, M. K. (1989). *Brain Res.*, **500**, 241.
28. Crespi, F., Garratt, J. C., Sleight, A. J., and Marsden, C. A. (1990). *Neuroscience*, **39**, 139.
29. Maidment, N. T., Martin, K. F., and Ford, A. P. D. W. (1990). In *Neuromethods* (ed. A. A. Boulton, G. B. Baker, and C. H. Vanderwolf), Vol. 14, p. 321. Humana Press, NY.
30. Gonon, F., Navarre, F., and Buda, M. (1984). *Analyt. Chem.*, **56**, 573.
31. Gonon, F., Buda, M., and Pujol, J. F. (1984). In *Measuring neurotransmitter release in vivo* (ed. C. A. Marsden), p. 153. J. Wiley & Sons, Chichester.
32. Gonon, F. G. and Buda, M. J. (1985). *Neuroscience*, **14**, 765.
33. Brazell, M. P., Kasser, A. J., Renner, K. J., Feng, J., Moghaddam, B., and Adams, R. N. (1987). *J. Neurosci. Meth.*, **22**, 167.
34. Cespuglio, R., Faradji, H., Hahn, Z., and Jouvet, M. (1984). In *Measurement of neurotransmitter release* in vivo (ed. C. A. Marsden), p. 173. J. Wiley & Sons, Chichester
35. Crespi, F., Paret, J., Keane, P. E., and Morre, M. (1984). *Neurosci. Lett.*, **52**, 159.
36. Sanghera, M. K., Crespi, F., Martin, K. F., Heal, D. J., Buckett, W. R., and Marsden, C. A. (1990). *Neuroscience*, **39**, 649.
37. Louilot, A., Buda, M., Gonon, F., Simon, H., LeMoal, M., and Pujol, J. F. (1985). *Neuroscience*, **14**, 775.
38. Stamford, J. A. (1990). *J. Neurosci. Meth.*, **34**, 67.
39. Millar, J. and Barnett, T. G. (1988). *J. Neurosci. Meth.*, **25**, 91.
40. Hafizi, S., Kruk, Z. L., and Stamford, J. A. (1990). *J. Neurosci. Meth.*, **33**, 41.
41. Richards, C. D. and Tegg, W. J. B. (1977). *Br. J. Pharmacol*, **59**, 526P.
42. Bull, D. R., Palij, P., Sheehan, M. J., Millar, J., Stamford, J. A., Kruk, Z. L., and Humphrey, P. P. A. (1990). *J. Neurosci. Meth.*, **32**, 37.
43. Ford, A. P. D. W. and Marsden, C. A. (1986). *Brain Res.*, **379**, 162.
44. Marsden, C. A., Joseph, M. H., Kruk, Z. L., Maidment, N. T., O'Neill, R. D., Schenk, J. O., and Stamford, J. A. (1989). *Neuroscience*, **25**, 389.
45. Maidment, N. T. and Marsden, C. A. (1987). *Neuropharmacology*, **36**, 187.
46. Maidment, N. T. and Marsden, C. A. (1987). *Eur. J. Pharmacol.*, **136**, 141.
47. Rivest, R., Jolicoeur, F. B., and Marsden, C. A. (1991). *Neuropharmacology*, **30**, 25.
48. Stamford, J. A., Kruk, Z. L., and Millar, J. (1988). *Brain Res.*, **454**, 282.
49. Stamford, J. A., Kruk, Z. L. and Millar, J. (1989). *Neuropharmacology*, **28**, 1383.
50. Leszczyszyn, D. J., Jankowski, J. A., Viveros, O. H., Diliberto, E. J., Near, J. A., and Wightman, R. M. (1991). *J. Neurochem*, **56**, 1855.

6

In vivo measurement of monoamine neurotransmitter release using brain microdialysis

TREVOR SHARP and TYRA ZETTERSTRÖM

1. Introduction

1.1 Biochemical measurements of monoaminergic neurotransmission

Not long after the discovery, in the 1950s, of mammalian brain neurons containing the monoamines noradrenaline (NA), dopamine (DA), and serotonin (5-hydroxytryptamine; 5-HT), biochemical methods were established to monitor the level of monoaminergic neurotransmission and to help us understand how this is altered by psychotropic drugs and the physiological state. These biochemical methods included the measurement, in animal brain tissue, of rates of monoamine synthesis, as well as levels of monoamine metabolites and their rate of clearance. At the same time methods were also developed to measure the efflux of monoamine neurotransmitters from nerve terminals in brain tissue *in vitro*.

A significant technical advance in this area was the development, beginning about 20 years ago, of methods capable of monitoring the release of mono-amine neurotransmitters in the brain *in vivo*. Such methods sample monoamine neurotransmitters, or their main metabolites, in the brain extracellular space, and include those which use implanted perfusion probes (for example push-pull perfusion) and voltammetric microelectrodes selective for either monoamines or their metabolites (1, 2).

1.2 Brain microdialysis

An exciting innovation in the field of neurotransmitter release measurement over the last few years has been the development of brain microdialysis (also termed intracerebral dialysis, intracranial dialysis, transtriatal/cortical dialysis). This is an *in vivo* brain perfusion method, originally used by Ungerstedt and his group at the beginning of the 1980s, to measure the release of rat striatal

DA (3, 4). In general terms this method operates like other types of intracerebral perfusion techniques. A physiological medium is applied, *in vivo*, to a discrete part of the brain tissue by passing it down into the brain via the inlet of a stereotaxically implanted perfusion probe. The flowing medium is exposed to the tissue at the probe tip and transmitters diffuse in from the extracellular fluid. The perfusate is then directed to the probe outlet where it is collected at regular intervals. By constantly monitoring transmitter levels in the perfusate, a localized measure of transmitter secretion in the brain tissue is obtained.

The great novelty of the microdialysis method is that the critical part of the perfusion probe, which exposes the perfusion medium to the tissue, is constructed from a semi-permeable dialysis membrane and, therefore, presents a *closed* perfusion system to the tissue. Unlike open-ended perfusion methods (for example push–pull perfusion), neurotransmitters in the brain extracellular fluid diffuse down the concentration gradient into the perfusing medium *via* the semi-permeable membrane. This simple innovation helps to preserve the integrity of the brain tissue around the implanted probe and avoids the constant washing of the tissue normally associated with open-ended perfusion probes. An important advantage of this is that the method can be applied to small laboratory animals, making the method more generally accessible to neuroscientists.

A further reason for the appeal of the microdialysis method is its versatility as a method for monitoring neurochemicals in the brain *in vivo*. Microdialysis perfusates contain detectable amounts of a large number of neuroactive substances, including amino acids, acetylcholine, purines, and various neuropeptides, as well as brain metabolites and ions (5, 6). Perfusate analysis is aided considerably by the fact that perfusates are relatively clean owing to large molecular weight substances in the extracellular space being excluded by the semi-permeable membrane.

The analysis of microdialysis perfusates using HPLC with electrochemical detection (EC), provides a method with the high sensitivity and selectivity necessary to detect the very low levels of *endogenous* monoamine neurotransmitters in the brain extracellular space. Previously it has proved very difficult to monitor the release of the monoamines without resorting to indirect methods, such as measurement of preloaded, radiolabelled monoamine or monoamine metabolite levels.

This chapter will describe in detail the microdialysis methodology currently used in this laboratory to measure *in vivo* extracellular levels of 5-HT, DA, and NA in the brain of the rat. Particular emphasis is given to the practicalities rather than the applications of the microdialysis technique, with the beginner very much in mind. Due to the restrictions of space this chapter will not cover, to any great depth, practical aspects of the HPLC–EC method (see ref. 7 for a review of HPLC–EC methodology). However, a sound working knowledge and hands-on experience of this method is essential to the successful application of the microdialysis technique for monoamine measurements. There is no substitute to gaining hands-on experience in a laboratory in which the microdialysis method is being used. A previous knowledge of stereotaxic surgery and HPLC–EC can, however, reduce the time spent learning the technique from months to a few weeks.

2. Microdialysis probes

There are three principal types of microdialysis probes; the single cannula probe, the short-loop probe, and the transcerebral probe (5, 8). The essential materials for constructing a microdialysis probe are:

- small-diameter semi-permeable dialysis tubing
- stainless steel tubing
- water-resistant, all purpose epoxy resin
- silica capillary tubing (see 2.1 below)

Details on the construction of the single cannula and short-loop microdialysis probes currently in use in this laboratory are described below. For information on the construction and implantation of the transcerebral probe the reader is referred to articles by the groups of Di Chiara (for example 5, 9) and Westerink (for example 10, 11).

2.1 Single cannula probes

The single cannula probe (*Figure 1*) is perhaps the most widely used type of microdialysis probe, and details of different designs (based on a similar theme) can be found in the literature (for example 12–15). A commercially available probe of this type is also available (Carnegie Medicin, via BAS and Biotech).

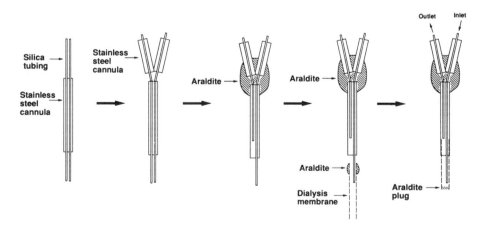

Figure 1. Illustration showing step-by-step construction of a single cannula type microdialysis probe. For further details see text (Section 2.1). Illustrations are not to scale.

The design of the probe described below (*Protocol 1*) is close to that previously outlined by Hutson *et al.* (16). The dimensions can be adjusted according to the size and depth of the target brain structure.

Protocol 1. Preparation of single cannula microdialysis probe

The following describes a step-by-step procedure for preparing the probe frame (see also *Figure 1*).

1. Insert two silica capillary tubes (25 μm o.d. vitreous silica, Scientific Glass Engineering, UK) cut to a length of about 30 mm using a scalpel blade, side-by-side into a 12 mm stainless steel cannula (0.5 mm o.d., wall thickness 0.06 mm, Goodfellows, UK) leaving about 10 mm exposed at one end (top of probe).

2. Attach a shorter steel cannula (6 mm) over each silica tube (top end of probe) to produce a fork-shaped frame. The two short cannulae eventually become the inlet and outlet to the probe.

3. Seal the junction of the main cannula and the two short cannulae by applying a drop of quick acting Araldite resin (Rapid Araldite; Ciba–Geigy). Before the resin is completely hard it is important that the outlet silica tubing is pulled about 5 mm into the main steel cannula.

4. Cut the inlet silica tubing to leave 2 mm exposed at the bottom end of the probe. Excess silica tubing at the top of the probe is trimmed with a scalpel blade.

Following preparation of the cannula frame the dialysis membrane is mounted.

5. Carefully apply a small amount of Araldite to the outer part of the end (1 mm) of a 12 mm length of tubular dialysis membrane (200 μm i.d., 40 000 MW cut-off, 60 Å pore size, Hospal AN 69). Pass the membrane over the silica inlet tube at the bottom of the probe, push 1–2 mm into the main cannula and leave to dry.

6. Cut the dialysis membrane using fine scissors to leave 3 mm exposed beyond the bottom of the main cannula. Finally, seal the open end of the dialysis membrane by lowering it into a small drop of Araldite, excess resin is wiped away from the tip using tissue paper. Store the probe in this form under dry conditions.

2.2 Loop-probes

Protocol 2. Preparation of loop-type microdialysis probe

The following describes a step-by-step guide to the construction of the loop-type microdialysis probe (see also *Figure 2*).

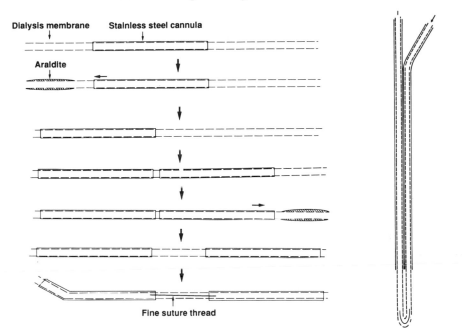

Figure 2. Step-by-step construction of a loop-type microdialysis probe (left). Completed probe on right. For further details see text (Section 2.2) Illustrations are not to scale.

1. Glue both ends (epoxy resin/Araldite) of a 5 cm length of dialysis tubing (cellulose, 270 μm o.d., Dow Corning, Michigan) inside stainless steel cannulae (18 × 0.4 mm) to leave approximately 5 mm of tubing exposed between the cannulae (the length of exposed membrane can be varied according to requirements). Additional glue can be applied to ensure that there is a good seal around the membrane at the ends of the cannulae. Take care not to touch the exposed membrane with moist fingers or epoxy resin.

2. Once the resin is dry, push a small piece of nylon filament (for example fine suture thread) into the lumen of the exposed part of the dialysis membrane using a fine tungsten wire. This prevents the dialysis tube from closing when folded into a loop. Bend one of the cannulae to an angle of about 45° to allow connection to the inlet tubing. At this stage the probe can be stored (under dry conditions).

The dialysis loop is formed on the day of experiment as follows.

3. Connect the inlet cannula to the perfusion system and, once the membrane is wet, bring the two cannulae together and bind tightly with a small piece of adhesive tape.

4. Mount the probe in the stereotaxic holder, remove the tape, and pass a tungsten wire down the lumen of the outlet cannula to carefully extend the membrane to form a tight loop. The tungsten wire is then temporarily fixed to the side of the holder using bone wax.

Protocol 2. *Continued*

5. The dialysis loop is now sufficiently rigid for implantation into the brain. Following implantation, remove the tungsten wire to allow connection of the outlet tube.

2.3 Properties of the dialysis membrane

The dialysis membrane is a critical component of the microdialysis probe since it interfaces the transfer of substances between the extracellular fluid and the perfusion medium. Dialysis membranes vary in permeability to monoamines, dimensions, molecular weight cut-off (~ 5000–$50\,000$), membrane composition (for example cellulose, polycarbonate, acrylic copolymer), and flexibility according to the manufacturer, and it is necessary to identify an appropriate membrane at an early stage.

From a practical point of view it is difficult to construct probes (in particular dialysis loops) from membranes that are brittle (acrylic copolymer) and narrow. In terms of membrane permeability to monamines, molecular weight cut-off of the membrane is not critical since the molecular weight of the monoamines is small in comparison. Permeability to monoamines, however, varies from membrane to membrane, probably due to small differences in wall thickness and/or the shape of the membrane pores rather than the membrane constituents. A measure of monoamine permeability can be determined by simple *in vitro* experiments (see Section 2.4 below). Intuition predicts that membranes with greater permeability to monoamines *in vitro* will recover proportionately greater amounts of monoamine from the brain *in vivo*, and indeed for many this has been the most important factor underlying membrane selection. However, this may be an over-simplification. It was recently reported that recovery (of monoamine metabolites) *in vitro* had a limited relationship to recovery *in vivo*; whilst membrane permeability to a particular solute has an important bearing on recovery *in vitro*, tissue resistance to solute movement is a more important determinant of recovery *in vivo* (17).

A major use of small-diameter dialysis tubing is as a vital component of renal dialysis machines. Bundles of dialysis tubing are fitted into replaceable cartridges; these are commercially available (manufacturers include Hospal Medical, Dow, Amicon, Enka; see ref. 18 for a full list of suppliers and addresses), one cartridge containing enough membrane for a lifetime's worth of experiments.

2.4 Assessing probe performance *in vitro*

When starting up the microdialysis method it is useful to test the performance of probes *in vitro* before moving on to the whole animal (*Protocol 3*). These experiments are simply carried out as follows.

Protocol 3. *In vitro* assessment of probe performance

1. Prepare the probe as described in *Protocol 1* or *2* and connect to perfusion system (see Section 3.1 below). Immerse the probe in a beaker of perfusion medium for about 15 min to thoroughly wet the dialysis membrane.

2. Transfer the probe to a beaker containing a solution of monoamine (10^{-5}–10^{-7} M made up in perfusion medium). The solution should be protected from light using silver foil and contain an antioxidant to prevent oxidation, for example 10^{-4} M ascorbic acid (see ref. 19).

3. Collect perfusates every 10–20 min and analyse along with the monoamine solution (at the beginning and end of the experiment to test for monoamine oxidation).

Using these procedures, the following basic properties can be demonstrated:

- The greater the surface area of the dialysis membrane, the greater the recovery of substances from the external medium (and the greater the chance of recovering detectable amounts of monoamine from the brain).

- The relative recovery (concentration in perfusate versus concentration in the outside medium) of substance through the membrane increases with decreasing flow-rate, equilibrium being reached at zero flow-rate (*Figure 3*). The absolute recovery of substance (amount of substance per unit time) is, however, relatively constant over the range 1–5 µl/min.

- The recovery of substance across the membrane is rapid, concentration-dependent and (obviously) directly related to the collection time (20, 21).

- The amount of substance recovered *in vitro* increases with increasing temperature (19).

In practical terms the *in vitro* experiments are particularly valuable for a number of reasons:

- they provide a test of the general durability and lifetime of the probe
- probe design, in particular membrane area, can be optimized to provide a high recovery of monoamine
- a measure of probe-to-probe variability can be obtained.

2.5 General comment: microdialysis probes and their operation

Flow-rates around 10 µl/min tend to lead to high pressure within the single cannula and loop-probe which not only results in increased incidence of failure of the sealed parts of the probe but may force perfusion medium through the pores of the dialysis membrane into the brain tissue. At flow-rates of less than

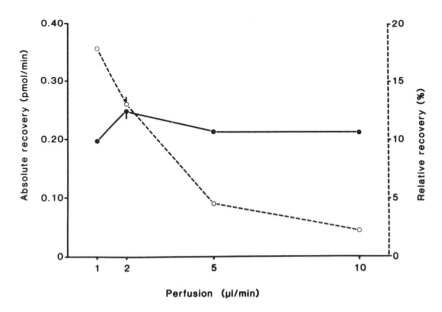

Figure 3. Graph showing typical data from an *in vitro* recovery experiment. In this case a loop-type dialysis probe was immersed in a solution containing 10^{-6} M dopamine. The relative (concentration of dopamine in perfusate versus concentration of dopamine in outside solution) and absolute (amount of dopamine per unit time) amounts of dopamine in the perfusates are plotted against perfusion rate (see Section 2.4).

$1\,\mu l/min$ probe blockage, usually due to air bubbles in the perfusion medium, tends to be more frequent and dead volume within the collecting tubes becomes a more critical factor. Unless there is a need to concentrate samples for assays requiring low volumes (for example microbore assays), low flow-rates of perfusion can be avoided. With the chronically implanted preparation there tends to be a greater incidence of probe failure due to factors such as leakage of the sealed joints, inlet/outlet blockage, and bent cannulae.

In practical terms the dialysis loop is easier to construct than the single cannula probe and has a relatively low failure rate. On the other hand, the dimension of the dialysing tip of the single cannula probe is about half that of the dialysis loop and thus the trauma and damage to the brain tissue associated with probe implantation is likely to be less. In addition, this makes the single cannula probe more useful for perfusing small brain nuclei. It should be added that both single cannula and short-loop microdialysis probes (and the transcerebral probe) are able to provide valid measurements of monoamine release as set out by the relevant criteria (see Section 6, *Table 3*).

3. Perfusion: equipment and perfusion media

3.1 Perfusion pump

Generally, microdialysis probes perform optimally when perfused at constant, low flow-rates in the range 1–5 µl/min. Small infusion pumps are better for this purpose than conventional peristaltic pumps. Several suitable pumps are available; this laboratory uses one supplied by Carnegie Medicin (CMA/100) (available in UK from Biotech) specifically built for microdialysis work with a number of useful features, for example fast forward/return modes, continuously adjustable flow-rate, and a microprocessor control option. Alternative pumps which have been found adequate for microdialysis work are the Harvard Microliter syringe pump and the Sage Model 341A pump.

The inlet to the dialysis probe is connected via polyethylene tubing (typically PE-20) to a reservoir of perfusion medium contained in a syringe (1 ml) mounted in the pump (up to nine 1 ml syringes can be mounted on the CMA/100). In anaesthetized animals, the perfusion pump is connected directly to the probe via polyethylene tubing. To perfuse awake animals it is necessary to incorporate a liquid swivel so that the animal can move freely within its environment without becoming entangled in the perfusion tubing (see also Section 4.3).

3.2 Switching perfusion media

Addition of drugs to the perfusion medium is a simple and useful way for administering drugs (via the dialysis membrane) directly into the brain and at the same time studying their effect on transmitter release. Switching between perfusion media is achieved simply and smoothly by manually disconnecting the polyethylene tubing supplying one medium (at a point close to the inlet of the dialysis probe so as to minimize the delay in drug entry to the probe) and rapidly reconnecting to another tubing supplying medium from a second syringe already mounted in the pump. Tubings can be coupled using a short piece of stainless steel tubing (hypodermic needle) which fits tightly in the tube lumen. Care should be taken not to introduce air bubbles during the switching procedure.

Carnegie Medicin have recently developed a fully automated perfusion media switching device (CMA/110/111). With this instrument it is possible to switch instantaneously between perfusion media using a motor-driven switching device with the option for microprocessor control.

3.3 Perfusate collection

Perfusates can be collected from the outlet of the dialysis probe in a number of ways. In the anaesthetized animal a simple approach is to cut off the tip of a small Eppendorf tube and invert it over a short length (1 cm) of polyethylene tubing connected to the probe outlet. Perfusate is held in the tip of the upturned tube by capillary retraction (see *Figure 1* in ref. 22). The collection tubes are replaced manually at each collection time point.

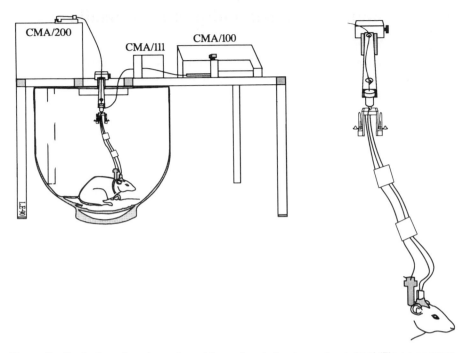

Figure 4. Illustration of equipment used for a chronic freely-moving animal (Figure courtesy of Carnegie Medicin, Stockholm).

For conscious animals, the collection tube needs to be secured to prevent spillage. It is common to collect from outlet tubing located at the level of the liquid swivel or (via the liquid swivel) outside the behavioural chamber (*Figure 4*). Alternatively, it is possible to mount a tube holder on a small rodent harness (see ref. 23 for illustrations and further details).

Although time-consuming, manual collection of perfusates allows one to utilize HPLC capacity to a maximum. With a collection time of 20 min and a sample run time of 7–8 min, a single HPLC can cope with output of perfusates from three preparations without a large backlog of samples. With manual perfusate collection it is also straightforward to split samples and analyse them separately.

As an alternative to manual collection of perfusates, an automated micro-fraction collector is available (Carnegie Medicin CMA/140). This device collects perfusate in glass vials held on a carousel which can be mounted directly into a refrigerated HPLC autoinjector (CMA/200). This takes away the need for constant supervision of the experiment, however, one loses the capacity to analyse samples as the experiment progresses.

A number of laboratories have taken the automation of collection further and directly coupled the outlet of the microdialysis probe to an automated load-valve on the HPLC thereby allowing on-line analysis of the perfusates (10, 21).

As with the automated sample collector, this method of collection and analysis commits an HPLC system to an individual animal but, nevertheless, fully automates the dialysis experiment.

3.4 Perfusion media

Our earlier experiments used, as a perfusion medium, a Ringer's solution comprising essential ions to maintain ionic balance. More recently we have adopted an artificial cerebrospinal fluid which is buffered and contains a greater complement of ions and glucose. The composition of this medium is:

- 140 mM NaCl
- 3 mM KCl
- 1.2–2.4 mM $CaCl_2$
- 1 mM $MgCl_2$
- 1.2 mM Na_2HPO_4
- 0.27 mM NaH_2PO_4
- 7.2 mM glucose, pH 7.4

The medium is prepared in distilled water as follows:

- 8.182 g/l NaCl
- 0.224 g/l KCl
- 0.203 g/l $MgCl_2$
- 0.43 g/l Na_2HPO_4
- 0.042 g/l NaH_2PO_4
- 1.297 g/l glucose
- 0.175 or 0.35 g/l $CaCl_2$ (slowly added at end to avoid precipitation)

3.5 General comment: perfusion media

An important feature of the microdialysis probe is that, unlike open-ended perfusion probes, there is no direct contact between the brain tissue and the perfusing media. Thus, it might be expected that the brain tissue in the vicinity of the microdialysis probe will remain bathed, to a large degree, in normal extracellular fluid rather than exogenous artificial media. The extent to which this is actually true is unknown. One would certainly expect any differences between the composition of the extracellular fluid and perfusion media to even out through equilibration across the dialysis membrane. However, homeostatic mechanisms within the brain may minimize disturbance to vital components of the brain extracellular environment. It should not be forgotten that since the exact content of the brain extracellular fluid is unknown, available perfusion media are approximations of the real thing.

At present a variety of perfusion media are utilized in microdialysis experiments and there is no consensus of opinion as to the most appropriate. There is, however, much evidence to suggest that even the simplest perfusion medium (for example Ringer's salt solution) allows the detection of monoamine which is released from neurons and changes in close accordance with monoaminergic neuronal activity (for example refs 9, 24, 25; see also Section 6).

3.5.1 Calcium levels

Most microdialysis studies have utilized perfusion media with calcium levels ranging from 1 to 3.4 mM. The extracellular concentration of free calcium in various rat brain regions is around 1–2 mM as detected by ion selective electrodes (26). A recent paper reported that perfusion with media containing 3.4 mM versus 1.2 mM calcium resulted in quantitative differences in drug-induced changes in DA (27). At this stage, although we know that the level of calcium in the perfusion medium clearly alters basal monoamine output (see Section 6.1), it is not clear to what extent it influences evoked changes in monoamine release. Where there are discrepancies in results between groups, however, differing calcium levels in the perfusion media may be a contributing factor.

3.5.2 Monoamine uptake blockers

A number of groups have found it necessary to add uptake inhibitors to the perfusion medium to measure endogenous 5-HT (using citalopram 1 μM) and NA (using desipramine 5 μM) on a routine basis (28–31). Whilst the presence of an uptake inhibitor may complicate certain pharmacological experiments, the localized inhibition of 5-HT or NA reuptake is likely to facilitate detection in the extracellular fluid of synaptically-released monoamine (experiments *in vitro* use uptake inhibitors for exactly this purpose); one can envisage situations when synaptic levels of transmitter increase but that these do not become detectable in the extracellular space due to rapid reuptake within the synapse. An example of this comes from recent experiments in which self-stimulating rats showed a release of DA in the nucleus accumbens which was only detectable in the presence of nomifensine, a DA reuptake inhibitor (32).

4. Microdialysis procedures

4.1 Implantation of microdialysis probes

The single cannula and short-loop dialysis probes are implanted into the brain tissue using standard stereotaxic techniques (*Protocols 4* and *5*). Necessary equipment are:

- surgical instruments
- a dental drill
- a stereotaxic frame (David Kopf)
- a bench microscope is extremely handy.

In addition a well-lit fume hood to isolate the animals and evacuate halothane and vapour from dental cement solvent is desirable.

Protocol 4. Preparation of microdialysis probe for implantation

1. Before commencing surgery, connect the microdialysis probes (which are stored under dry conditions) to the perfusion system and mount in a suitable stereotaxic holder.

2. Clamp the holder in a small retort stand and lower the probe into a beaker containing perfusion medium to allow thorough wetting and a slight expansion of the dialysis membrane. Once wet the membrane should not be left out of solution for more than a few minutes since drying can result in deterioration in recovery of monoamine across the membrane.

3. At this point inspect the probe carefully to ensure that it is free of blockage of flow and leaks.

4.1.1 Anaesthesia

In non-recovery experiments (acute preparation), we typically anaesthetize rats using either chloral hydrate (450 mg/kg i.p.; 100 mg chloral hydrate/ml saline injected in a volume 4.5 ml/kg) or halothane (0.15% halothane/air mixture). Halothane is preferred, as an anaesthetic, to prepare a chronic preparation since recovery is quick and usually without incident. Halothane is added to the air source (supplied by a small air pump) using a vaporizer and administered to the animal via a face mask.

4.1.2 Surgical procedures

Protocol 5. Intracerebral implantation of microdialysis probe

1. Mount the anaesthetized animal in the stereotaxic frame and expose the skull by a single 1 cm incision. Clean and dry as far as possible. Access to the skull is aided by the use of forceps to withdraw the two folds of skin to each side of the head.

2. Drill burr holes (~2 mm diameter) for implantation of the dialysis probe(s) taking care to leave the dura membrane intact. Then mount one or two stainless steel screws into additional burr holes. These screws will serve as an anchor point for the dental cement which will follow.

3. Next, mount the microdialysis probe (prepared as in *Protocol 4*) on the stereotaxic frame and position according to its rostrocaudal and mediolateral stereotaxic coordinates (relative to bregma) and lower to make contact with the dura surface (as viewed by microscope) to zero the dorsoventral stereotaxic co-ordinate. Lift the probe off the dura which is then broken

159

Protocol 5. *Continued*

with the point of a hypodermic needle, taking care not to unnecessarily rupture blood vessels on the cerebral surface.

4. Before implantation of the probe, wash and thoroughly dry the skull and cerebral surface. The latter step is important since dental cement will not adhere to wet surfaces. Furthermore, occasionally we have encountered abnormally high levels of 5-HT (>0.15 pmol/20 min) which persist throughout the experiment. This contamination is probably a result of contact of the dialysis membrane with blood during the implantation procedure.

5. Implant the dialysis probe slowly into the brain tissue (>1 mm/min). This is to prevent damage to the delicate tip of the probe and limit disruption to the tissue. Fast implantation of the probe may lead to a continuous rise in perfusate levels of the catecholamine metabolites DOPAC and HVA, as well as a progressive fall of DA, over the several hours. It is highly likely that a too rapid implantation of the probe increases the chance of contaminating perfusates with transmitter from damaged neurons and non-neuronal sources.

6. Once in position, dry the skull a final time before carefully pouring a fast-drying dental cement on to the skull and around the probe and skull screws. As the cement is drying, a useful tip is to release the forceps and allow the skin to close in around the probe. Further cement can then be applied to both the cement base and the top of the wound, thereby providing a neat and firm seal and avoiding the need to use sutures.

7. Once the cement has hardened, remove the probe holder and strengthen the mount with further cement to ensure that the shaft of the probe is well supported.

8. The preparation is now ready for the acute experiment. For the chronic preparation take particular care to ensure that both the cement mount and seal to the wound are sound and that the height of the mount and the length of exposed probe are kept to a minimum. Finally, cut the polyethylene tubes connecting to the inlet and outlet of the probe (leaving 5–10 mm attached to the probe) and seal their ends with bone wax. Once recovered from anaesthesia, return the animal to its home cage.

4.2 Acute (anaesthesized) microdialysis experiments

In these experiments animals are maintained under anaesthesia throughout the experiment. We have routinely used both halothane and chloral hydrate as the anaesthetic for these experiments. Halothane (0.15% halothane/air mixture) has the advantage that it allows maintenance of a relatively constant and long-lasting level of anaesthesia. However, the need for a vaporizer and air supply for each animal makes halothane relatively cumbersome to administer compared to systemically administered anaesthetics such as chloral hydrate. In the latter case, supplementary doses need to be administered throughout the course of the experiment. With choral hydrate (i.p. administration) one can achieve a stable level of anaesthesia over at least eight hours using a dose rate of approximately 60 mg/kg/h (dosing intervals tend to vary between animals).

We have not observed a consistent effect of chloral hydrate injection on dialysate levels of either 5-HT or NA during the course of an experiment. However, to minimize drug interactions or spurious results, as a rule of thumb we avoid injecting chloral hydrate one hour either side of the acute injection of a drug. Routinely, body temperature is maintained at around 35–36 °C using either a thermoregulated heating pad or a lamp (60 W bulb) positioned over the animal.

In terms of time course, a typical experiment to test the effect of a drug on monoamine release in the acute (anaesthetized) preparation would run as follows: surgery (30 min), then a 3 h baseline period (perfusates collected every 20 min), drug injection followed by a 2–3 h post-drug period. The long baseline period is critical as discussed below (Section 4.4).

4.3 Chronic (freely moving) microdialysis experiments

For experiments on monoamine release in relation to the behavioural or physiological state of the animal there is little option but to use the freely moving preparation. As well as equipment such as a liquid swivel and a means for perfusate collection which allows the animal to be perfused with minimal restriction to its movement (see Section 3.4), the other important accessory is a suitable animal chamber in which the relevant behavioural measurements can be obtained. The simplest chamber is a hemispherical bowl with the liquid swivel placed centrally overhead. A commercially available version of this is illustrated in *Figure 4*. We have previously described a Perspex activity chamber based on a circular running track contained within a sound-resistant air-conditioned chamber (23). Four horizontal photocell beams linked to a digital readout counter are placed symmetrically around the track and provide a sensitive measure of locomotor movement.

A number of experimental set-ups for chronic microdialysis experiments are reported in detail in the literature (for example 21, 33). Also described are replaceable microdialysis probe systems which allow probes to be inserted into guide cannula which had been previously implanted (16, 21).

4.4 General comment: acute versus chronic microdialysis experiments

4.4.1 Tissue damage

Acute intracerebral implantation of small cannulae or probes leads immediately to a number of disturbances in the vicinity, including spreading depression, damage to the blood–brain barrier, and disrupted glucose metabolism which

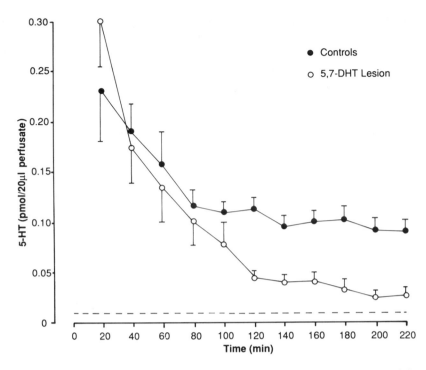

Figure 5. 5-HT in hippocampal microdialysates collected over time immediately following implantation of a short-loop probe into control and 5-HT-denervated rats. The dotted line represents the approximate limit of detection. Note that 5-HT levels in the lesioned group fall below controls about 2 h post-implantation. Data taken from ref. 29 with permission.

reverse within minutes or hours (6). These disturbances plus the inevitable neuronal damage resulting from probe implantation probably account for the high levels of monoamine measured in perfusates during the first hour or so following implantation of the microdialysis probe (*Figure 5*). However, there is evidence to suggest that the greater the time elapsed between probe implantation and the commencement of the monoamine measurements, the more negligible is the amount of the transmitter derived from damaged neurons or non-neuronal sources. For example, the level of 5-HT in hippocampal dialysates from rats with lesioned serotonergic neurons only falls below that of non-lesioned controls after about two hours of perfusion (*Figure 5*; ref. 29). This suggests that only later in the experiment (>2 h) does 5-HT in hippocampal dialysates (at least measured under our experimental conditions) derive principally from serotonergic neurons.

Therefore, providing the basal output of monoamine has reduced and been stable over about two hours, the acute (anaesthetized) microdialysis experiment should be able to provide a valid measure of monoamine release, and this will be evident in terms of the appropriate response of monoamine output to certain pharmacological and physiological stimuli (see Section 6). When there is doubt as to whether monoamine output measured in the acute preparation satisfies the relevant criteria it is advisable to allow the brain tissue longer to recover from the trauma of probe implantation and carry out the experiment 24 h post-surgery.

4.4.2 Lifetime of the chronic preparation

It is now well recognized that within a few days of implantation of the microdialysis probe a gliotic reaction develops in the tissue immediately surrounding the dialysis membrane; the result is a coating of the membrane with inflammatory cells and connective tissue which forms a barrier to diffusion (cf. ref. 6). This probably accounts for the finding that both the basal and drug-evoked output of monoamine in microdialysates begins to decline within a few days of probe implantation (*Figure 6*; refs 10, 30, 34). Nevertheless, basal output of both striatal DA at 4 days and hippocampal 5-HT at 7 days is reported to be Ca^{2+}-dependent (30, 34) suggesting that both are of neuronal origin even at this time.

The changing condition of the tissue and downward drift of monoamine output over time following probe implantation clearly complicates studies of treatment effects on monoamine release over time on the same animal. This problem, however, is likely to be common to all techniques monitoring monoamine release using intracerebral probes.

4.4.3 Use of anaesthesia

An important consideration when opting for the anaesthetized preparation is whether anaesthesia will compromise the value of the experiment. In the case of halothane, for example, there is clear evidence that it can alter *quantitatively* the response of DA output to DA antagonists (cf. refs 35, 36). Since in this particular case the anaesthetic does not *qualitatively* alter the DA response, the value of the experiment with halothane seems not to be in doubt, but then it can only be judged on the basis of the experiment carried out in the absence of anaesthesia.

Given that experiments using microdialysis with awake animals are technically more trying and time-consuming, one has to ask whether by using anaesthesia one can arrive at the same answer in half the time using less animals and with minimal discomfort to them. When in doubt the answer probably is to carry out preliminary experiments with and without anaesthesia.

5. Measurement of monoamines and metabolites using HPLC–EC

5.1 Basic principles

The HPLC assay utilizing a carbon-based electrochemical detector electrode measures catechol and indole compounds with high sensitivity and selectivity (7),

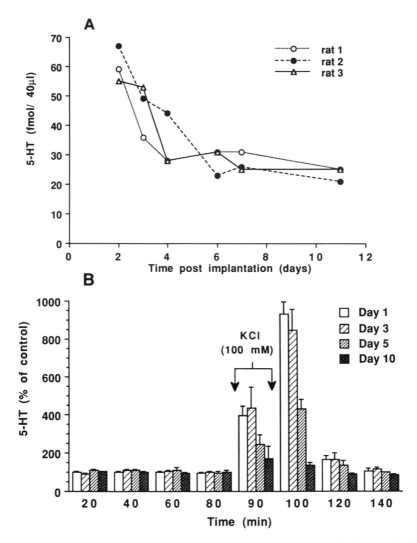

Figure 6. Figure showing basal (a) and 100 mM potassium-stimulated (b) levels of 5-HT in hippocampal microdialysates collected from the awake freely-moving rat over several days following probe implantation. Note the gradual decline in both basal and evoked output of 5-HT. Data taken from ref. 30, with permission.

and for many is currently the method of choice for the routine analysis of monoamines in brain microdialysates. Measurements can be made by HPLC–EC without sample extraction or derivatization and are usually sufficiently rapid to allow perfusates to be analysed continuously as the experiment progresses. This latter facility offers an important advantage over, for example radioenzymatic or mass spectrometric assays, not least because drugs can be

Table 1. Examples of HPLC–EC assays used to detect monoamines and their metabolites in rat brain dialysates

Mobile phase	HPLC column	Separation mode	Substances analysed	Reference
0.1 M sodium acetate pH 4.1, 1.8 mM heptane sulphonate, 0.1 mM EDTA, 6–8% methanol	Nucleosil, C_{18}, 5 μm	reversed phase-ion pair	DA DOPAC, HVA	10
0.23 M sodium acetate, 0.015 M citric acid, 100 mg/l EDTA	Supelcosil, C_{18}, 5 μm	reversed phase	DA DOPAC, HVA	9
0.05 M acetate–citrate, pH 5.2	Nucleosil SA, 10 μm	cation exchange	DA	20
0.03 M sodium citrate, 0.078 M sodium acetate, pH 4.7, 6.5% methanol	Supelcosil LC-8-DB, C_{18} 5 μm	reversed phase	5-HT, 5-HIAA	25
0.026 M phosphoric acid pH 2.6, 0.065 mM sodium octanyl sulphonate, 1% acetonitrile, 3% tetrahydofuran	Supelcosil LC-8-DB, C_{18} 3 μm	reversed phase-ion pair	5-HT, 5-HIAA	46[a]
0.1 M sodium phosphate, 1 mM sodium octanyl sulphonate, 0.1 mM EDTA, 9% methanol	Brownlee Velosep RP-18, 3 μm	reversed phase-ion pair	NA DOPAC, 5-HIAA	40
0.1 M sodium dihydrogen phosphate pH 4.5, 0.1 mM sodium octanyl sulphonate, 0.1 mM EDTA, 9% methanol	Spherisorb 5ODS2, C_{18}, 5 μm	reversed phase-ion pair	NA, DA adrenaline, DOPAC, 5-HIAA, MHPG	48

[a]fluorimetric detection

administered in the knowledge of stable baseline values. Examples of HPLC–EC assays which have been successfully applied to detect monoamine neurotransmitters in brain microdialysates are shown in *Table 1*. Details on the assays that we are currently using are given in Section 5.3.

A thorough knowledge of the practicalities of HPLC, in particular optimization of chromatographic separation and detector sensitivity, is essential in order to apply it in combination with the dialysis technique. One is faced

Table 2 Typical basal levels of dopamine, 5-HT and NA in microdialysates collected under a variety of experimental conditions

Monoamine	Level[a] (pmol/20 min sample)	Brain region	Probe type	Anaesthesia[b]	Uptake blockers[c]
DA	0.05 ± 0.008 (6)	striatum	DL	yes	no
	0.080 ± 0.008 (5)	striatum	SC	no	no
	0.039 ± 0.005 (5)	n. accumbens	SC	no	no
5-HT	0.009 ± 0.002 (7)	hippocampus	DL	yes	no
	0.060 ± 0.003 (5)	hippocampus	DL	yes	yes
	0.080 ± 0.01 (6)	hypothalamus	DL	yes	yes
	0.060 ± 0.004 (6)	frontal cortex	DL	yes	yes
	0.051 ± 0.004 (6)	hippocampus	SC	yes	yes
	0.040 ± 0.006 (4)	hippocampus	SC	yes	yes
NA	0.02 ± 0.006 (4)	hippocampus	DL	yes	no
	0.09 ± 0.010 (5)	hippocampus	DL	yes	yes
	0.065 ± 0.005 (6)	hippocampus	SC	yes	yes

DL—dialysis loop; SC—single cannula probe
[a] Data obtained from a typical group, mean ± SEM (n = number of rats) values
[b] Chloral hydrate
[c] 5-HT—citalopram 1 µM, noradrenaline–desipramine 5 µM

not only with the problem of resolving and detecting very small amounts of monoamine in the microdialysates (*Table 2*), but also with the fact that chromatograms of microdialysates are invariably littered with rogue, and often unidentified, peaks which need to be resolved from those of interest. In addition, whilst it should be quite possible to set up an HPLC assay from a published method, small adjustments to the mobile phase may be necessary to cope with batch-to-batch differences between columns, column age, sample contaminants, etc.

One of the most widely used separation modes for monoamines and their acid metabolites involves the use of reversed-phase ion-pair chromatography conditions. Generally speaking this combines a conventional HPLC column (150–250 mm length, 4.6 mm i.d.; packed with C_{18} 5 μm particles) with a mobile phase comprising an aqueous buffer (for example phosphate, acetate–citrate), an ion pair reagent (for example sodium octanyl sulphonate) and an organic solvent (for example methanol, tetrahydrofuran). Below is a rule of thumb guide as to how changes in certain components of the mobile phase alter the retention time of monoamines and their acid metabolites (DOPAC, HVA, and 5-HIAA):

- An increase in pH decreases the retention time of acid metabolites but has little effect on the monoamines.
- An increase in concentration of detergent ion-pairing agents (for example sodium octanyl sulphonate) increases the retention time of the monoamines but changes that of the acid metabolites little.
- An increase in the amount of organic solvent decreases the retention time of both monoamines and their metabolites.

It should be emphasized that this is only a guide; other factors, in particular ionic strength of the mobile phase, column length, column packing, and temperature, also determine retention time of monoamines and their metabolites (see ref. 37 and refs therein).

5.2 HPLC equipment

The following is a list of HPLC equipment currently in use in this laboratory:

- LKB 2150 HPLC pump
- Rheodyne 7125 injector (20 or 50 μl loop)
- Bioanalytical Systems BAS LC-4B electrochemical detector
- Milton Roy CI-4000 Integrator.

We have utilized HPLC columns (150–250 × 4.6 mm) packed with a variety of materials (for example Spherisorb 5ODS2 from Chrompak; Ultrasphere ODS from Beckman; C_{18} 3 and 5 μm particles). Currently, as a cost-saving move, we equip our assays with columns with replaceable cartridges (Rainin Instruments Company, USA).

5.3 Chromatography conditions for separation of monoamines and their metabolites

Below are chromatography conditions optimized to allow separation of DA, 5-HT, or NA in regional rat brain microdialysates; in each case the monoamine acid metabolites DOPAC and 5-HIAA are present on the chromatogram. The mobile phase for individual assays comprises the same components but in different proportions. The retention time for the monoamine in each case is around 5 minutes, which allows adequate separation from the solvent front without a significant loss of resolution due to peak broadening. Despite much effort we have not yet been able to establish conditions which allow the detection in control microdialysates of any two monoamine neurotransmitters simultaneously: samples have to be divided and analysed on separate systems.

All of the mobile phases below are prepared with purified water and filtered before use through a 0.45 μm membrane (Millipore). Mobile phase is prepared in batches of 5 litre and stored at room temperature in stoppered flasks.

5.3.1 Dopamine (DA)

Mobile phase
- 0.12 M phosphate
- 15% (v/v) methanol
- 0.5 mM sodium octanyl sulphonate
- 0.1 mM EDTA
- pH 3.8

Preparation (5 litre):
- 600 ml of 1 M $NaH_2PO_4 \cdot 2H_2O$ stock (pH to 3.0 using phosphoric acid)
- 3400 ml water
- 14.9 ml of 10% (w/v) EDTA stock
- 0.322 g sodium octanyl sulphonate (Sigma)
- 710 ml HPLC grade methanol
- adjust pH to 3.8 using concentrated sodium hydroxide

Chromatography column
150 × 4.6 mm packed with Dynamax Microsorb C_{18} 5 μm particles (supplied by Rainin). Flow-rate; 1–1.2 ml/min.

Detector cell
BAS TL-17A thin-layer glassy carbon cell (25 μm gasket) maintained at +0.65 to +7.0 V versus a silver/silver chloride electrode.

A typical chromatogram is shown in *Figure 7*.

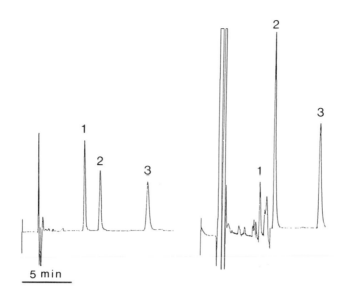

Figure 7. Chromatograms (integrator plots) from HPLC–EC assay optimized to detect dopamine in rat brain dialysates. **Peak 1**: DA, **peak 2**: DOPAC, **peak 3**: 5-HIAA (not shown HVA which elutes after 16–18 min). Chromatogram on left is of 50 µl of a 1 pmol standard solution. Right is a chromatogram of 50 µl of dialysate collected from the nucleus accumbens of a freely-moving rat; DA, 0.05 pmol; DOPAC, 17.8 pmol; and 5-HIAA, 11.7 pmol.

5.3.2 Serotonin (5-HT)

Mobile phase

- 0.12 M phosphate
- 12.5% (v/v) methanol
- 0.01 mM sodium octanyl sulphonate
- 0.1 mM EDTA
- pH 3.8.

Preparation (approx. 5 litre):

- 600 ml of 1 M $NaH_2PO_4 \cdot 2H_2O$ stock (pH to 3.0 using phosphoric acid)
- 3520 ml water
- 14.9 ml of 10% (w/v) EDTA stock
- 0.022 g sodium octanyl sulphonate
- 590 ml HPLC grade methanol
- adjust pH to 3.8 using concentrated sodium hydroxide

169

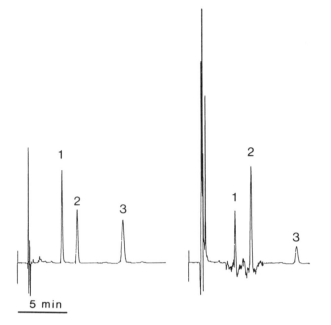

Figure 8. Chromatograms (integrator plots) from HPLC–EC assay optimized to detect 5-HT in rat brain dialysates. **Peak 1:** 5-HT, **peak 2:** DOPAC, **peak 3:** 5-HIAA. Chromatogram on **left** is of 20 μl of a 2 pmol standard solution. **Right** is a chromatogram of 20 μl dialysate collected from the ventral hippocampus of an anaesthesized rat: 5-HT, 0.04 pmol; DOPAC, 0.11 pmol; 5-HIAA, 1.53 pmol.

HPLC column and detector as for DA

A typical chromatogram is shown in *Figure 8.*

5.3.3 Noradrenaline (NA)

Mobile phase
- 0.1 M phosphate
- 12% (v/v) methanol
- 2.0 mM sodium octanyl sulphonate
- 0.1 mM EDTA
- pH 4.6

Preparation (approx. 5 litre):
- 500 ml of 1 M $NaH_2PO_4 \cdot 2H_2O$ stock (pH to 3.0 using phosphoric acid)
- 3900 ml water
- 14.9 ml of 10% (w/v) EDTA stock
- 2.16 g sodium octanyl sulphonate

170

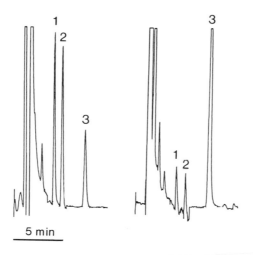

Figure 9. Chromatograms (integrator plots) from HPLC–EC assay optimized to detect noradrenalin in rat brain dialysates. **Peak 1:** DOPAC, **peak 2:** noradrenaline, **peak 3:** 5-HIAA. Chromatogram on **left** is of 20 µl of a 2 pmol standard solution. **Right** is a chromatogram of 50 µl dialysate collected from the ventral hippocampus of an anaesthetized rat. DOPAC, 0.12 pmol; noradrenaline, 0.12 pmol; 5-HIAA, 6.4 pmol.

- 600 ml HPLC grade methanol
- adjust pH to 4.6 using concentrated sodium hydroxide

HPLC column and detector as for DA
A typical chromatogram is shown in *Figure 9*.

5.4 Standards and sample storage

Stock solutions containing a mixture of standards at a concentration of 10^{-3} M are made up in 0.1 M perchloric acid and stored in a refrigerator at 4 °C for up to 3 months. This stock solution is diluted to 10^{-6} M each week and a 10^{-7} M standard is prepared from this each day for injection on to the assay.

Microdialysate samples are injected directly on to the HPLC immediately following collection whenever possible and always on the day of the experiment (if this is not the case the stability of the monoamines under storage needs to be verified). On occasions samples may build up and can be stored on ice before injection.

6. Method validation

Bearing in mind the invasive nature of the microdialysis technique (see Section 4.4), it is important to ensure that monoamines in the brain micro-dialysates arise from functional, rather than damaged, monoaminergic neurons or non-neuronal sources. A number of validatory experiments need to be carried

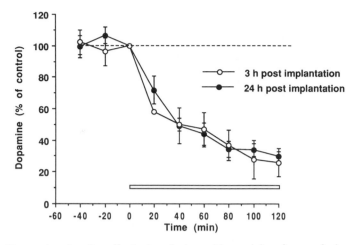

Figure 10. Figure showing the effect of perfusion with a calcium-free perfusion medium on dopamine levels in microdialysates collected from the striatum of the chloral hydrate anaesthetized rat, 3 and 24 hours following implantation of a short-loop microdialysis probe. The calcium-free perfusion medium was applied as indicated by the horizontal bar. Note that dopamine output falls markedly and to a similar extent in both preparations.

out in order to cover for this possibility (and at the same time help confirm the identity of the nature of the peak on the chromatogram). Selected, important experiments of this type are given below; for further details the reader is referred to the original articles and several recent reviews on this topic (5, 11, 38).

6.1 Sensitivity to Ca^{2+} and tetrodotoxin

A number of microdialysis studies have demonstrated that basal output of DA (9, 11), NA (39, 40), and 5-HT (25, 29, 31) falls markedly when the perfusion medium is changed to one with calcium omitted or in which the sodium channel blocker tetrodotoxin is added, as expected from what we know about the physiological processes underlying neurotransmitter release.

In order to determine whether monoamine output is Ca^{2+} and/or tetrodotoxin-dependent the following experiments should be carried out.

Ca^{2+} dependency

Make up a batch of perfusion medium as in Section 3.5 (with uptake blocker if appropriate) and a second batch with $CaCl_2$ omitted. Perfuse the animal with standard perfusion medium for 2–3 h and obtain a stable output of monoamine. On switching to the Ca^{2+}-omitted perfusion medium, levels of monoamine should decline rapidly (see *Figure 10*); after 60–80 min switching to the standard perfusion medium should result in a return of monoamine levels towards control levels. The reduction should be at least 50%, although this will depend on how close basal monoamine levels are to the limit of detection of the assay.

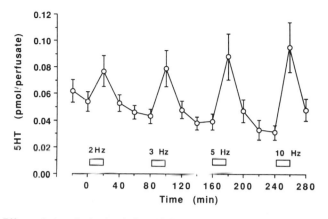

Figure 11. Effect of electrical stimulation of the dorsal raphe nucleus on 5-HT levels in the ventral hippocampus of the chloral hydrate anaesthetized rat. Note the frequency-dependent increase in 5-HT output. Data taken from reference 47 with permission.

Sensitivity to tetrodotoxin

Perfuse the acute or chronically prepared animal with standard perfusion medium for 2–3 h and obtain a stable output of monoamine. Switch to perfusion medium containing 1 or 10 μM tetrodotoxin for 60 min and then return to standard perfusion medium. During perfusion with tetrodotoxin monoamine levels should fall markedly and then, on its removal, recover towards normal.

6.2 Physiological stimulation

Neuronal depolarization will evoke a Ca^{2+}-dependent increase of monoamine levels into the microdialysis perfusates. Depolarization can be induced as follows:

- *High potassium*—Following stabilization of monoamine output, switching to perfusion medium containing high potassium (50–100 mM) for 20 min will evoke a marked and short-lasting increase in monoamine output. Potassium should have little effect on monoamine output if Ca^{2+} is omitted from the perfusion medium prior to the application of high potassium (Ca^{2+}-omitted) medium

- *Nerve stimulation*—Electrical stimulation of monoaminergic cell body regions or axon bundles in the mid-brain evokes the predicted increase in microdialysate levels of NA (39), DA (9) and 5-HT (*Figure 11*; refs 29, 47). The methodology involved in this type of experiment is relatively complex and the reader is referred to the original papers.

Our experience is greatest with hippocampal 5-HT, evoking an increase in output by electrically stimulating the serotonergic neurons in the dorsal raphe nucleus (*Figure 11*; ref. 29, 47). In brief the protocol is as follows:

Protocol 6. Electrical stimulation of 5-HT output in hippocampus

Experiments are carried out on the chloral hydrate anaesthetized rat. Perfusion medium is as described in Section 3.5 containing 2.4 mM Ca^{2+} and 1 µM citalopram. Perfusates are collected every 20 min.

1. Shortly after implantation of the dialysis probe into the ventral hippocampus (see *Protocol 5*), lower a stimulating electrode (co-axial bipolar stainless steel electrode; Clark Electrical Instruments, Pangbourne, UK) into the dorsal raphe nucleus where it remains for the rest of the experiment.

2. Connect the electrode via a Grass isolation unit (PS1U6) to a Grass stimulator (S48). Stimulate as follows: 1 ms cathodal monophasic pulses, 300 µA current, 2–10 Hz pulse frequency and 20 min train duration.

3. In all experiments allow a baseline period of 140–180 min post-dialysis probe implantation before commencing electrical stimulation. We have applied two types of experimental design:

   ● *Frequency response paradigm.* Apply stimulations for 20 min every 60 min with stimulation frequency increasing step-wise from 2 Hz to 10 Hz during the course of the experiment. In our experience a small (+20%), but reproducible, increase in 5-HT output is obtained with a pulse frequency of 2 Hz. Where this response is not obtained it is usually due to incorrect placement of the electrode; lowering the electrode to a new position should resolve the problem.

   ● *Twin pulse paradigm.* In this paradigm, apply two 20 min periods of stimulation at 3 Hz separated by a 100–120 min control period. A similar amount of 5-HT is released in each stimulation period. Drugs can be added between the periods of stimulation, the first period of stimulation serving as a paired control for the second.

6.3 Autoreceptor agonists/antagonists

It is well known that monoamine release in the brain can be altered by drugs which act on inhibitory autoreceptors localized on both the monoaminergic nerve terminal and somatodendrites. Once the microdialysis method is established in the laboratory it is relatively straightforward to test whether monoamine levels in brain microdialysates decrease and increase in response to the administration of monoamine agonists and antagonists, respectively (for DA: see refs 9, 20, 41 and 42; NA: see refs 39 and 40; 5-HT: see refs 28 and 43).

In this type of experiment drugs are injected systemically after the 2–3 h baseline period. Responses to monoamine agonists and antagonists are rapid in onset and depending on dose, last for several hours. Subcutaneous injection of the monoamine receptor agonists apomorphine (0.05–0.5 mg/kg),

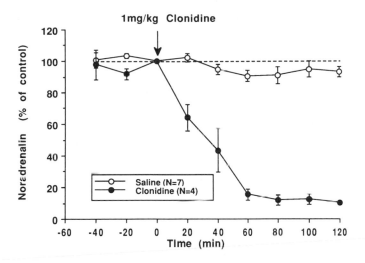

Figure 12. Effect of injection of the α_2 agonist clonidine (1 mg/kg s.c.) on noradrenaline levels in the ventral hippocampus of the chloral hydrate anaesthetized rat. Note the rapid and marked drop in noradrenaline output. Data from C. Done, this laboratory.

clonidine (0.1–1 mg/kg) and 8-OH-DPAT (0.025–0.25 mg/kg) will effectively reduce perfusate levels of DA, NA and 5-HT, respectively, and these effects are reversed by their respective antagonists sulpiride (50 mg/kg), idazoxan (0.5 mg/kg), and pindolol (8 mg/kg). Although this type of experiment is relatively straightforward, it is strongly recommended that the original studies are consulted before embarking on work of this nature. *Figure 12* shows the effect of clonidine upon the NA content of hippocampal dialysates.

6.4 Neuronal lesions

Selective and extensive neurotoxic lesion of monoaminergic neurons should lead to a marked decrease in monoamine output. This has shown to be the case for DA (44), NA (39, 45), and 5-HT (29, 46; see *Figure 5*).

Lesioning of monoaminergic neurons should be carried out using well-characterized procedures described in the literature. It is important to validate the lesion biochemically or histochemically before applying it to the microdialysis experiment. A further point to note is that animals with only partial lesions may show normal levels of transmitter output since neurons spared by the lesion increase in activity to compensate for the overall loss of transmission. Clearly, as a test of the source of monoamine in the perfusates, the lesion experiment is not as straightforward as the other methods described in 6.1 and 6.3 above.

6.5 General comment: validity of different experimental conditions

The above criteria have been met for a variety of experimental conditions, in particular different microdialysis probes and acute and chronic preparations (see *Table 3*). However, there are a small number of experiments reported in the literature where it is evident that monoamines in the brain microdialysates did not arise from functional monoaminergic neurons. In particular, experiments using microdialysis probes of the loop-type acutely implanted in striatum have been described in which the output of DA was not Ca^{2+}-dependent (10) and the output of 5-HT was not diminished by 5-HT neuronal lesions (46). Whilst these findings suggest that the acute preparation might not meet the above criteria, data from a number of studies suggest that this is likely to be the exception rather than the rule (*Table 3*). These discrepancies, nevertheless, emphasize the need to validate one's own experimental conditions; a crucial factor for the acute preparation would appear to be an extended baseline period before commencement of the transmitter measurements (see *Section 4.4.1*).

7. Concluding remarks

The availability of biochemical techniques to monitor the release of monoamines in the animal brain *in vivo* is vital for a fuller understanding of how

Table 3 Effect of pharmacological and physiological manipulations on monoamine neurotransmitters in brain microdialysates

Monoamine	Preparation[a]	Ca^{2+}/TTX sensitivity	Neuronal lesion	Agonists/ antagonists	Physiological stimulation
DA	acute	DL[5e] SC (34) TC (9)	DL (45)	DL (20, 22)	TC (9)[b]
	chronic	DL[5] SC (34) TC (10)		TC (5)	
5-HT	acute	DL (29, 46)	DL (29)	DL (28)	DL (29, 46)[b]
	chronic	DL (44) TC (25)	DL (44)	DL (44) TC (25)	DL (30)[c]
NA	acute	DL[d] TC (39)	DL (38)	DL[d] TC (39)	DL[b,d] TC (39)[b]
	chronic	DL (31)	DL (48)	DL (40)	DL (40)[c]

DL—dialysis loop; SC—single cannula probe; TC—transcerebral probe; TTX—tetrodotoxin
[a]*Acute*—on day of probe implantation, *Chronic*—at least 24 h post-probe implantation.
[b]Electrical stimulation of the mid-brain.
[c]Tail-pinch.
[d]C. Done, this laboratory, unpublished observation.
[e]This chapter, see *Figure 10*.
Reference numbers in parentheses.

monoaminergic neurotransmission is altered by psychotropic drugs and physiological state. These techniques allow us to take into account the complexities of monoamine neuronal function, particularly as they relate to interactions between the monoamines and other neurotransmitters, neuroanatomical sites of drug action in the brain, and the relationship between monoamine release and behaviour. The development of microdialysis as a means for monitoring the release of DA, 5-HT, and NA in rat brain is already playing an ever increasing and important role in this area, as are the voltammetric techniques described in Chapter 5.

For practical purposes the basic principles of the microdialysis technique are straightforward and the skills required to run the method are well within the scope of research students, experienced technicians, and scientists alike. The essential equipment does not require major investment, much of it being standard in modern neuropharmacological laboratories with a strong interest in brain monoamine function.

This chapter has described in detail the methodology based on ongoing experiments in this laboratory, that should allow the measurements to be carried out on a routine basis. There is little doubt that this methodology will advance. Having to detect very low amounts of the monoamines in the extracellular space creates the need for collection times in the order of several minutes at least and, in certain circumstances, the use of uptake inhibitors to boost spillover of the monoamine from the synapse. We have already applied an HPLC–EC assay for DA based on microbore columns to reduce sample collection times to one minute (49); perfusate analysis of monoamines using this method should become commonplace now that such columns are being produced commercially to a high standard. Future efforts to miniaturize the microdialysis probe to tip diameters less than 100–200 µm face the problem of coping with high perfusion pressure and will put further demands on the sensitivity of the analytical method.

It is evident from the surge in the number of groups applying the microdialysis technique over the last few years that many consider its present weaknesses a small price to pay for its strengths as a tool for investigating monoamine transmission *in vivo*.

Acknowledgements

We are grateful to Steve Bramwell, Christoper Done, Qi Pei, Richard Mcquade, and Sasha Gartside for their intellectual and technical input into the experiments underlying this paper. We thank Lena Ekwall (Carnegie Medicin) for the supply of figures.

References

1. Myers, R. D. and Knott, P. J. (ed.) (1986). *Neurochemical analysis of the conscious brain: voltammetry and push–pull perfusion*. Vol. 473. Ann. N.Y. Acad. Sci.

2. Marsden, C. A., Joseph M. H., Kruk Z. L., Maidment N. T., O'Neill R. D., Schenk J. O., and Stamford J. A. (1988). *Neuroscience*, **25**, 389.
3. Ungerstedt, U., Herrara-Marschitz, M., Jungnelius, U., Ståhle, L., Tossman, U., and Zetterström, T. (1982). In *Advances in dopamine research; advances in the biosciences* (eds M. Kohsaka *et al.*), pp. 219–231. Pergamon Press, New York.
4. Zetterström, T., Sharp, T., Marsden, C. A., and Ungerstedt, U. (1983). *J. Neurochem.*, **41**, 1769.
5. Di Chiara, G. (1990). *Trends Pharmacol. Sci.*, **11**, 116.
6. Benveniste, H. (1989). *J. Neurochem.*, **52**, 1667.
7. Marsden, C. A. and Joseph, M. H. (1986). In *HPLC of small molecules: a practical approach* (ed. C. K. Lim), pp. 29–48. Practical Approach Series, IRL Press, Oxford.
8. Ungerstedt, U. (1984). In *Measurement of neurotransmitter release* in vivo, (ed. C. A. Marsden). *IBRO Handbook Series, Vol. 6*, pp. 81–107. John Wiley, Chichester.
9. Imperato, A. and Di Chiara, G. (1984). *J. Neurosci.*, **4**, 966.
10. Westerink, B. H. C. and Tuinte, M. H. J. (1986). *J. Neurochem.*, **46**, 181.
11. Westerink, B. H. C., Damsma, G., Rollema, H., De Vries, J. B., and Horn, A. S. (1987). *Life Sci.*, **41**, 1763.
12. Johnson, R. D. and Justice, J. B. (1983). *Brain Res. Bull.*, **10**, 567.
13. Sandberg, M. and Lindström, S. *J. Neurosci. Meth.*, **9**, 65.
14. Clemens, J. A. and Phebus, L. A. (1984). *Life Sci.*, **35**, 671.
15. Robinson, T. E. and Whishaw, I. Q. (1988). *Brain Res.*, **450**, 209.
16. Hutson, P. H., Sarna, G. S., Kantamaneni, B. D., and Curzon, G. (1985). *J. Neurochem.*, **44**, 1266.
17. Hsaiao, J. K., Ball, B. A., Morrison, P. F., Mefford, I. N., and Bungay, P. M. (1990). *J. Neurochem.*, **54**, 1449.
18. Kendrick, K. M. (1989). In *Neuroendocrine peptide methodology* (ed. P. M. Conn), pp. 229–252. Academic Press, New York.
19. Parry, T. J., Carter, T. L., and McElligott, J. G. (1990). *J. Neurosci. Meth.*, **32**, 175.
20. Zetterström, T., Sharp, T., and Ungerstedt, U. (1984). *Eur J. Pharmacol.*, **106**, 27.
21. Church, W. H. and Justice, J. B. (1987). *Analyt. Chem.*, **59**, 512.
22. Strecker, R. E., Sharp, T., Brundin, P., Zetterström, T., Ungerstedt, U., and Björklund, A. (1987). *Neuroscience*, **22**, 169.
23. Sharp, T., Ljungberg, T., Zetterström, T., and Ungerstedt, U. (1986). *Pharmacol. Biochem. Behav.*, **24**, 1755.
24. Westerink, B. H. C., Hofsteede, H. M., Damsma, G., and De Vries, J. B. (1988). *Naunyn-Schmeideberg's Arch. Pharmacol.*, **337**, 373.
25. Carboni, E. and Di Chiara, G. (1989). *Neuroscience*, **32**, 637.
26. Silver, I. A. and Erecinska, M. (1990). *J. Gen. Physiol.*, **95**, 837–866.
27. Moghaddam, B. and Bunney, B. S. (1989). *J Neurochem.*, **53**, 652.
28. Sharp, T., Bramwell, S. B., and Grahame-Smith, D. G. (1989). *Brit. J. Pharmacol.*, **96**, 283.
29. Sharp, T., Bramwell, S. B., and Grahame-Smith, D. G. (1989). *J. Neurochem.*, **53**, 234.
30. Pei, Q., Zetterström, T., and Fillenz, M. (1989). *Neurochem. Int.*, **15**, 503.
31. Kalen, P., Nilson, O. G., Cenci, M. A., Rosengren, E., Lindvall, O., and Björklund, A. (1989). *J. Neurosci. Meth.*, **34**, 107.
32. Nakahara, D., Ozaki, N., Kapoor, V., and Nagatsu, T. (1989). *Neurosci. Lett.*, **104**, 136.
33. Kuczenski, R. and Segal, D. (1989). *J. Neurosci.*, **9**, 2051.

34. Osborne, P. G., O'Connor, W. T., Drew, K. L., and Ungerstedt, U. (1990). *J. Neurosci. Meth.*, **34**, 99.
35. Zetterström, T., Sharp, T., and Ungerstedt, U. (1986). *Naunyn-Schmiedeberg's Arch. Pharmacol.*, **334**, 117.
36. Ståhle, L., Collin, A.-K., and Ungerstedt, U. (1990). *Naunyn-Schmeideberg's Arch. Pharmacol.*, **342**, 136.
37. Kontur, P., Dawson, R., and Monjan, A. (1984). *J. Neurosci. Meth.*, **11**, 5.
38. Marsden, C. A. (1990). In *The pharmacology of noradrenaline in the central nervous system* (eds D. J. Heal and C. A. Marsden) pp. 155–186. Oxford University Press, Oxford.
39. L'Heureux, R., Dennis, T., Curet, O., and Scatton, B. (1986). *J. Neurochem.*, **46**, 1794.
40. Abercrombie, E. D., Keller, R. W., and Zigmond, M. J. (1988). *Neuroscience*, **27**, 897.
41. Zetterström, T. and Ungerstedt, U. (1984). *Eur. J. Pharmacol.*, **97**, 29.
42. Imperato, A. and Di Chiara, G. (1988). *J. Pharmacol. Exp. Ther.*, **244**, 257.
43. Sharp, T. and Hjorth, S. (1990). *J. Neurosci. Meth.*, **34**, 83.
44. Zetterström, T., Brundin, P., Cage, F. H., Sharp, T., Isacson, O., Dunnett, S. B., and Björklund, A. (1986). *Brain Res.*, **362**, 344.
45. Kalen, P., Kokaia, M., Lindvall, O., and Björklund, A. (1989). *Brain Res.*, **474**, 374.
46. Kalen, P., Strecker, R. E., Rosengren, E., and Björklund, A. (1988). *J. Neurochem.*, **51**, 1422.
47. Sharp, T., Bramwell, S. R., and Grahame-Smith, D. G. (1990). *Neuroscience*, **39**, 629.
48. Routledge, C. and Marsden, C. A. (1987). *Neuroscience*, **20**, 457.
49. Carlsson, A., Sharp, T., Zetterström, T., and Ungerstedt, U. (1986). *J. Chromatog.*, **368**, 299.

7

Antibody microprobes

A. W. DUGGAN

1. Introduction

The antibody microprobe (1, 2) was developed as a means of getting better spatial precision in determining sites of release of neuropeptides in the central nervous system with minimal disturbance to the structures giving release. As such it overcomes some of the difficulties associated with other techniques used to measure release. These include the inability of surface perfusion methods to localize sites of release with any accuracy, and the trauma associated with the introduction of steel cannulae or flexible tubing into the brain and spinal cord.

Glass microelectrodes are among the least disruptive devices for introduction into the central nervous system. There is currently no method which gives a satisfactory on-line signal from microelectrodes adequate for detecting the minute concentrations of the multiplicity of neuropeptides present in the brain and spinal cord. What the microprobe does is immobilize antibodies, to a neuropeptide of interest, on to the outer surfaces of glass microelectrodes. When placed in the central nervous system, if a peripheral stimulus causes a release of neuropeptides, a proportion of released molecules will bind to that part of the microprobe adjacent to the site of release. The microprobe is then withdrawn and incubated in a radiolabelled form of the neuropeptide under study. The focally bound, neurally-released molecules are detected on autoradiographs as deficits in the binding of the labelled peptide. The principle is shown in *Figure 1*.

2. The preparation of antibody microprobes

2.1 Glass and the pulling of glass

Since the glass has to be drawn by heating in a microelectrode puller, Pyrex tubing alone has been used. Normally the shanks of glass microelectrodes have an abrupt taper as this gives a lower resistance for a given tip size. However, such a shape is not suitable for microprobes since these ultimately have their tips broken off for placement on X-ray film. For this to be done without shattering the tip (and losing an unknown length) a very flexible microprobe is required and, hence, one drawn with a very gradual taper. Typical dimensions are shown in *Figure 2*.

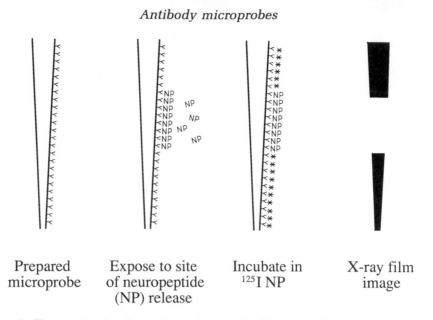

| Prepared microprobe | Expose to site of neuropeptide (NP) release | Incubate in ^{125}I NP | X-ray film image |

Figure 1. The principle of antibody microprobes. For simplicity, the antibodies are only shown on one side of a glass micropipette.

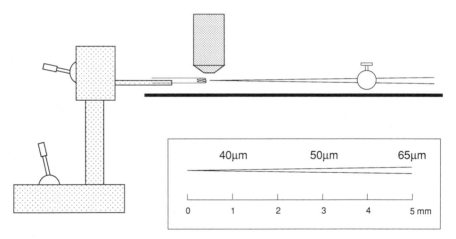

Figure 2. The proportions of a typical microprobe tip (inset) and the micromanipulator arrangement used for heat sealing the tip.

This sort of taper is most readily achieved by starting with relatively large diameter Pyrex glass tubing (outer diameter 3 mm, internal diameter 2 mm). The puller used by the author has a heating coil extending over 8 mm of the Pyrex tubing thus ensuring the melting of a large area of glass and hence producing micropipettes with a gradual taper. The pulled micropipettes are heat-sealed at both ends since antibody coating of the outside alone is needed.

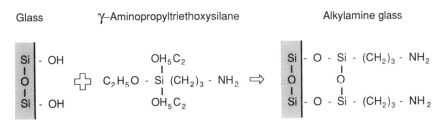

Figure 3. The interaction between γ-aminopropyltriethoxysilane and a glass surface. The substituted silane is shown as interacting with silanol groups, but a reaction with adsorbed water is probably also important.

The unpulled end is sealed in a gas flame. The tip is sealed under microscopic control by touching it against an electrically-heated coil, mounted in a micromanipulator. It is easier to do this by having a small glass bead on a winding of the heating coil and touching this bead when molten rather than the bare wire. *Figure 2* shows the arrangement of heating coil and microprobe used for tip sealing.

2.2 Placing organic groups on micropipettes

Organic groups suitable for binding proteins can be attached to glass using organosilicon compounds (3–5). With antibody microprobes γ-aminopropyltriethoxysilane (APTES) is used as the starting compound. The initial reaction is shown schematically in *Figure 3*. The hydroxyl groups interacting with the substituted silane are shown as part of the glass, but there is evidence that much of the reaction depends on chemisorbed water molecules. Provided there is sufficient adsorbed water on the surface of the glass, a siloxane polymer will build up. A polymeric as opposed to a monomeric coating is crucial for the success of the microprobe technique. Microprobes seek to localize sites of release in the central nervous system with a resolution of 50–100 µm, and for this to be possible it is essential that the coating of antibodies is even. Currently the best way to ensure this is for the initial silane–glass reaction to result in a polymer which is visible under a light microscope.

The description of the method used for microprobe coating which follows is based on having as little water vapour as possible in the solution of the substituted silane, in which micropipettes are immersed, and a controlled amount of adsorbed water present on the surfaces of micropipettes. Prior to adopting the above, the success rate with coating micropipettes was very dependent on the weather!

2.3 The substituted silane

Although the method uses γ-aminopropyltriethoxysilane (APTES) (6) other substituted silanes have been experimented with but with inferior or unimproved results. Considerable care must be taken to decant the purchased APTES (Aldrich) into small boiling tubes devoid of excess water vapour (*Protocol 1*).

Protocol 1. Decanting of γ-aminopropyltriethoxysilane (APTES)

1. A vacuum oven with an upward opening chamber is needed. A conventional vacuum oven placed on its end with the door opening upwards is used in the author's laboratory.

2. Place 10–20, 20 ml Quickfit tubes into a beaker so that they stand upright. Place this beaker, tube stoppers in a glass bowl, together with the barrel and plunger of a 50 ml glass syringe and a long 12 gauge needle in the vacuum oven.

3. Dry overnight.

4. Next morning switch off the oven and as it cools (normally taking at least 2 hours) introduce dry argon into the chamber to discharge the vacuum.

5. Open the vacuum chamber and, while continuing to introduce argon (which will be retained because of the vertical orientation of the chamber) place the unopened bottle of APTES in the chamber. Rubber gloves should be worn and the room should be well ventilated. Inhaled APTES will liberate ethanol in the air passages and, while not dangerous, this is not pleasant. Having the vacuum chamber vertical also minimizes escape of APTES vapour.

6. Remove the top of the APTES bottle, assemble the glass syringe and needle, and decant 10 ml of APTES into each of the small boiling tubes. Stopper the tubes. A small piece of Teflon tape around each stopper helps prevent sticking. The aliquots of APTES are then stored in a domestic refrigerator.

2.4 Siloxane coating of microprobes

The details of coating sealed glass micropipettes with a visible siloxane polymer are given in *Protocol 2*.

Protocol 2. Coating the microprobes

1. Take 10–15 heat-sealed micropipettes and place them in a glass carrier (*Figure 4*). These carriers have three perforations in the base to permit draining during the many washing procedures used during microprobe preparation. Immerse the microprobes and glass carrier in 10% nitric acid for 30 minutes. Wash 3 times for 5 minutes in distilled water and once in Millipore-filtered distilled water. Then suspend the carrier by its hooked upper end in a dedicated clean oven and dry for 2 hours at 200 °C.

2. Take another glass carrier and fill it with molecular sieves (Aldrich 4A, 4–8 mesh) and place the carrier in the oven used to dry microprobes.

3. Place a clean boiling tube (Quickfit 24/29, large enough to take the microprobe carrier but small enough to fit within the centrifuge buckets to be used subsequently), in the same oven as the microprobes and dry for 2 hours.

4. Remove the boiling tube to a fume cupboard and, while still warm, introduce 40 ml of reagent grade toluene and 10 ml of previously aliquoted APTES. Stopper the solution and place it in a rack.

5. Remove the carrier containing the molecular sieves from the oven, allow to cool for approximately one minute and then plunge them into the APTES–toluene solution. There will be some boiling of the latter. Replace the stopper as soon as possible.

6. Transfer the APTES–toluene tube containing molecular sieves in a carrier to the inside of a box saturated with water vapour at room temperature (18–20 °C). A small (non-functioning) refrigerator is used in the author's laboratory.

7. With metal forceps remove the carrier containing the micropipettes and suspend it within the water-saturated box. Close the doors of the latter and after one minute open the box, remove the top of one APTES–toluene containing boiling tube, remove the carrier containing molecular sieves then plunge the carrier containing microprobes into the APTES–toluene solution. Since the carrier is still hot there will be some boiling of the mixture and hence eye protection is advisable. The author has performed this procedure several hundred times without any splashing of the APTES–toluene solution. This procedure has not been carried out in a fume cupboard, but instead a well-ventilated room is used.

8. Stopper the boiling tube immediately and place in a centrifuge bucket. Normally four carriers of microprobes are treated together and they are then centrifuged at 6 °C, 2000 r.p.m. for one hour. The object of centrifugation is to displace free water away from the tips of microprobes.

9. Remove the stoppered tubes from the centrifuge, place them in a holder and allow to stand for 12 hours at room temperature.

10. Remove each carrier and rinse once in toluene. Each carrier batch of microprobes is then inspected under an incident light, binocular microscope for adequacy of coating. If this appears satisfactory the microprobes are cured, still within the carrier, in a dedicated oven at 200 °C for 24 hours.

If the coating reaction has gone satisfactorily, the APTES–toluene solution will still appear clear after 12 hours. The coated microprobes will appear to have a uniform milky appearance when viewed under incident light using a magnification of ×40. This is illustrated in *Figure 4*. Following heat curing,

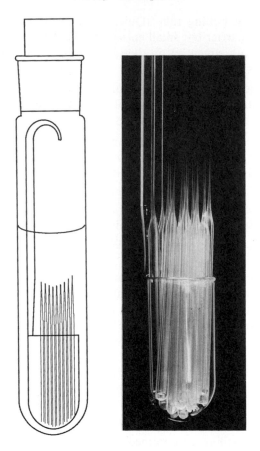

Figure 4. Microprobes and carriers. The diagram on the left shows microprobes in a glass carrier and immersed in the APTES–toluene solution. The photograph to the right shows actual microprobes following the initial coating procedure. The faint milky appearance of the microprobe tips contrasts with the clear glass of a new glass carrier.

each microprobe is checked for evenness of coat under transmitted light, the appearances of medium and heavy coats are shown in *Figure 5*. This coat should resist removal by wiping, done by holding a microprobe in one hand and a piece of soft tissue around the shank with the thumb and forefinger of the dominant (other) hand and then pulling the probe through the tissue. No matter how hard one squeezes with the fingers of the hand holding the tissue, a good coating will not be visibly altered by this procedure. Slight irregular lumps in the coating may be removed by this wiping procedure.

Unsatisfactory coating is usually associated with a clumpiness in the APTES–toluene solution at the end of the incubation time. It is probable that a polymerization has occurred in the bulk of the solution and the resultant polymer

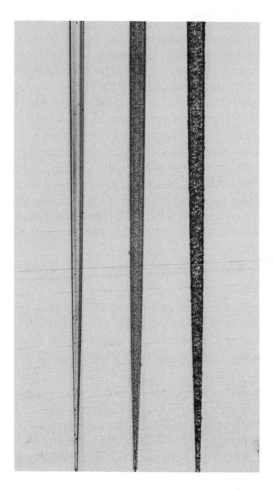

Figure 5. Photomicrographs showing, from left to right, untreated glass, a light coating and a heavy coating of siloxane polymer. Although both coatings are adequate for *in vivo* use, it would be better to adopt one rather than use microprobes with very different coatings.

settles on microprobes and is not bonded. Such probes, when inspected under incident light microscopy, are often coated irregularly with fluffy clumps which easily wipe off even after heat curing. Such probes need not be discarded. They should not be heat cured but rather wiped clean and the coating procedure repeated. Unsatisfactory coating probably most commonly results from an excess of water (in the APTES–toluene solution).

Microprobes are stored flat in the author's laboratory, each lying in a groove cut in a metal bar held within a flat plastic box.

2.5 Adding antibodies

It takes 3 days for coated microprobes to be made ready for an *in vivo* experiment. The technical details are described in *Protocol 3*. Prior to undertaking the practical side there are some theoretical considerations. Proteins are immobilized to microprobe surfaces via the bifunctional reagent glutaraldehyde. If a monoclonal antibody concentrate virtually free of unwanted proteins is available, then the antibodies could be coupled directly from such a concentrate. However, in practice antisera raised from animal immunization are usually used, and, hence, some means of immobilizing immunoglobulins is needed. This is done by means of protein A or protein G. Protein A is suitable for antisera raised in rabbits and mice but will not bind the immunoglobulins present in antisera from sheep and goats (7). Protein G will bind not only rabbit, sheep, and goat immunoglobulins but also other proteins including albumin. The newer genetically-engineered, protein G is relatively free of this disadvantage of binding albumin.

Protocol 3. Adding antibodies

1. Take the required number of heat-cured, siloxane-coated microprobes and place them in glass carriers of the type previously used for coating. Submerge them in a small measuring cylinder containing 2.5% solution of glutaraldehyde for 30 minutes and then wash 3 times for 5 minutes in distilled water.

2. Obtain an aliquot of protein A (or protein G, where appropriate) and dilute in PBS, with sodium azide 0.1%, to a concentration of 0.1 mg/ml. Fill a number of 5 µl glass capillaries with the protein A solution, and under a binocular, incident-light microscope insert a glutaraldehyde-treated microprobe into each capillary. This is most readily done by using a magnification of ×30–40 and holding the capillary in the non-dominant hand at an angle of 30–40° to the horizontal so that the top opening is clearly in focus. Bring the microprobe up with the dominant hand and, as it enters the capillary, slacken the hold of the latter with the other hand. If the capillary is held rigidly, a lack of parallel alignment of probe and capillary (which is virtually unavoidable) will result in the microprobe being broken.

3. Place each microprobe, capped with a capillary, in a slot on a Perspex rack. When all microprobes are treated, place the rack in a flat shallow tray containing water-soaked paper on the base, cover the tray with a Perspex or metal cover, and place in a cold room for 24 hours.

4. After incubation remove the protein A-containing capillaries from the probes. The capillaries will usually fall if the probes are held vertically with tips down. Place each microprobe in a large Petri dish containing sodium borohydride 2.5% w/v in a borate buffer. This is to reduce Schiff bases produced by aldehyde coupling to amino groups of protein A. Hydrogen

188

bubbles will form both in the solution and on the microprobes. Therefore, lift up the tips of the latter transiently with forceps, one at a time, so that the bubbles burst and the probe is then replaced. **Caution**: this process should be carried out in a fume cupboard and flames are to be avoided. After 10 minutes in borohydride, replace the microprobes in glass carriers and wash 3 times in PBS for 5 minutes.

5. Dilute an aliquot of antibody in PBS–azide and fill 5 or 25 µl capillaries with this solution. Insert the microprobes in the capillaries as previously described and incubate in a cold room for 24–48 hours depending on the dilution of antibody. After incubation, remove the capillaries, place the microprobes in glass carriers containing a solution of PBS. This is normally done on the day of an *in vivo* experiment. Inspect each microprobe with transmitted-light microscopy and number them.

Both protein A and protein G can be purchased fairly pure and hence can occupy all of the available sites on microprobes if desired. Theoretically, although there may be free aldehyde sites after the binding of protein A or protein G, these have not been a problem. Experiments which used either ethanolamine or glycine to block such sites showed no differences in the performance of microprobes treated this way from that of untreated probes.

In a conventional radioimmunoassay it is usual to decide both an appropriate antibody dilution and an amount of tracer to result in the assay measuring within a range appropriate to the amounts present in the sample being measured. Within limits, sensitivity is improved by reducing the amount of tracer and by increasing the dilution of the antibody. With microprobes this has not been done. The procedure described in *Protocol 3* aims to have the maximum amount of antibody present on microprobe surfaces. It has proved satisfactory for use in the central nervous system and can provide answers within 3–5 days of an *in vivo* experiment.

Antibodies purchased commercially or obtained from colleagues come in a variety of forms. Some commercial sources even lyophilize the antibody extensively diluted in buffer so that one receives a large bottle of salts! However, none of these differences are serious when it is remembered that antibody concentration is commonly performed by means of protein A immobilized to a column solid support. Microprobes bearing protein A are essentially analogous to such columns. Thus when setting up microprobes and using an antiserum for the first time, the dilution of the antiserum is not that critical. If the concentration of antibodies is particularly low (for example when using antibodies contained in a commercial radioimmunoassay kit) then use a large capillary for the antibody incubation and allow a longer time. If a particularly sensitive assay is required, and hence a lesser amount of antibody is appropriate on microprobe surfaces, then this is best done by adding varying amounts of albumin to the protein A solution so reducing the number of antibody binding sites. Such microprobes, however, will require X-ray film exposures of several weeks.

3. Tests prior to *in vivo* use

Prior to performing *in vivo* experiments, it is necessary to test the sensitivity of microprobes, the non-specific binding of the radiolabelled ligand, and the specificity of the immobilized antibody.

3.1 Measurements of *in vitro* sensitivity and non-specific binding

This involves incubating microprobes in known concentrations of the peptide being studied, with subsequent processing being identical to that employed after use in the central nervous system. With nearly all of the microprobe types used to date it has been shown that preincubation in the synthetic ligands at a concentration of 10^{-8} M for 30 min at 23 °C has suppressed the binding of the labelled ligand by $> 50\%$ (1). The procedure for *in vitro* assay is outlined in *Protocol 4*.

Protocol 4. *In vitro* 'calibration' of microprobes

1. Take 24 microprobes from the batch prepared for an *in vivo* experiment.
2. Prepare 100 µl aliquots of the neuropeptide being studied in concentrations of 10^{-8}, 10^{-7}, and 10^{-6} M.
3. From each aliquot fill six 5 µl capillaries. Insert a marked microprobe into each capillary and place in a humidified incubator at 37 °C for 30 minutes.
4. Remove from the incubator and wash the probes for 15 minutes in ice-cold stirred PBS–Tween solution.
5. Place all of the microprobes, including those not exposed to unlabelled neuropeptide, in capillaries containing labelled neuropeptide as described previously.
6. Incubate and wash as with *in vivo* microprobes (*Protocol 5*).
7. Glue the broken off tips to separate small pieces of cardboard and place them in counter tubes.
8. Count the radioactivity of each probe.
9. If desired, obtain autoradiographs of these microprobes. They can be glued to the lower part of the paper bearing *in vivo* microprobes.

The amounts of labelled ligand, displaced by prior occupancy of endogenous neuropeptide, which can be detected on autoradiographs of microprobes are extremely small; for example: ^{125}I—Bolton–Hunter substance P with specific activity of approximately 2000 Ci/mmol results in 5×10^{18} d.p.m./mol (1 Ci being 2.22×10^{12} d.p.m.). With an average microprobe bearing antibodies to substance P, incubation in the tracer results in bound radioactivity of approximately 150 d.p.m./mm of length. Since complete inhibition of binding over 100 µm can be readily detected, this corresponds to 3×10^{-18} moles of non-bound labelled substance P. Although the relationship between the numbers of bound endogenous molecules and excluded tracer molecules is unknown, there can be little doubt that the amount of bound endogenous ligand is extremely small.

The amounts bound, however, are really of little physiological value. What is needed is an estimate of the concentrations in the vicinity which produce a given inhibition of binding. Even this is only of real value for a compound which is degraded slowly and, hence, diffuses widely. With a rapidly degraded compound, what is detected is determined by the amount released and the rate of degradation, the latter usually being unknown. By making comparisons with *in vitro* assays it is possible, using microprobes, to make estimates of the concentrations of neuropeptide in the vicinity.

3.2 Tests of the specificity of the antibody

Although the antisera used are examined by conventional radioimmunoassays, it is also important to use *in vitro* assays with microprobes in examining antibody specificity. To date no discrepancies have emerged between the two types of assay. Peptides of close structural similarity to the one being measured should be tested.

Currently it does not appear possible to characterize the ligand bound to a microprobe by chemical means. Even making generous assumptions about complete inhibition of such binding over a zone of 100 µm does not result in estimates of bound ligand which are measurable physicochemically. Thus the identity of what is bound can only be rendered highly probable by obtaining similar results with different antibodies recognizing different sequences of the one polypeptide. In one such series of experiments microprobes bearing antibodies either to the N-terminus or the C-terminus of substance P gave essentially similar results when examining spinal release due to the development of peripheral arthritis. It should be noted that with antibodies to the N-terminus of a neuropeptide, the use of Bolton–Hunter iodinated peptide is contra-indicated. In the experiments of Schaible *et al.* ^{125}I–Tyr^8–substance P was used as the tracer (8).

4. Use in the central nervous system

4.1 Placing the microprobes into the brain and spinal cord

On the morning of an experiment, microprobes are removed from antibody-containing capillaries, inspected under a transmission light microscope for

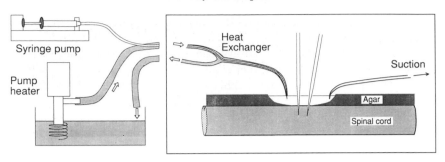

Figure 6. A diagram of the *in vivo* use of antibody microprobes. Not shown are the micromanipulators used to introduce microprobes.

possible damage to the coating, numbered with a felt-tip pen, and stored (in carriers) in PBS. They are introduced into the brain or spinal cord in the same way as normal microelectrodes. In the author's laboratory, two are introduced at a time using stepping motor-driven micromanipulators. They are nearly all filled with a solution of pontamine sky blue 2% (w/v) in sodium acetate since this enables the tips to be easily seen, permits extracellular recording of the firing of neurons, and the ejection of dye from the tip for subsequent location in histological sections. Filling with dye entails first removing the sealed blunt ends by scratching with a glass-cutting knife and breaking off between the thumbs and forefingers of the two hands. The initial scratching must be fairly deep so that the pressure exerted in breaking off the end is minimal. If excessive pressure is used, an unpredictable length of the tip end will also be lost and the microprobe may be unusable. The sealed tip end is broken off, using the manipulator-controlled coil originally used for sealing but without heating. The dye solution is then introduced via finely drawn out polythene tubing attached to a Pasteur pipette. If the microprobe does not immediately fill pressure should be applied to the blunt end using a large syringe and tubing.

With microprobes having tip diameters of 5–10 µm, the pia mater of the adult cat cannot be penetrated. Thus it is necessary to remove small areas of pia mater (usually approximately 0.2 mm in diameter) at sites of proposed penetration. This raises the question of sterility and studies of neuropeptide release. Inflammatory exudates contain neuropeptides and proteases. For both reasons inflammatory exudates should be minimized or excluded from studies employing antibody microprobes. In experiments on the spinal cord of the cat this entails:

- opening the dura mater with sterile forceps and scissors
- covering the spinal cord with sterile Ringer–agar
- removing part of the agar cover at sites of proposed microprobe penetration with sterile forceps
- removing small amounts of pia mater
- irrigating the exposed area of spinal cord with sterile Ringer's solution

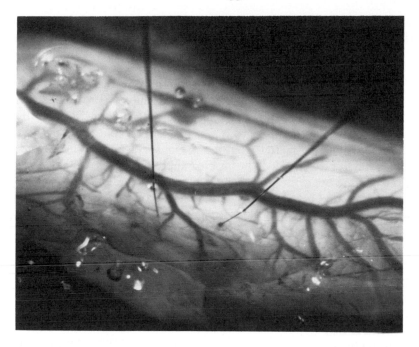

Figure 7. Two microprobes inserted 3 mm into the lumbar spinal cord of the cat. The microprobes are filled with pontamine sky blue. The agar covering of the spinal cord has been removed around the site of penetration and is filled with Ringer's solution giving refractive bending of the lower parts of the microprobes.

Irrigation normally uses an infusion pump driving the sterile Ringer's through a tube enclosed in a heat exchanger for nearly all of its course to the spinal cord. A fine drawn out polythene piece delivers the sterile Ringer solution to the spinal cord surface.

Figure 6 illustrates the arrangement used, and *Figure 7* shows two dye-filled microprobes in place in the spinal cord.

It is not the object of this account to advise on *in vivo* experimental protocols but some aspects do require comment. It needs to be emphasized that release implies increased levels following a defined stimulus. Thus it is necessary to employ a number of microprobes as controls for the effects of the stimulus in each experiment. For microprobes to detect release they have to be in the right place. A stimulus to a limited site (for example to the skin of hind limb digits) will result in release of compounds in a restricted area. Microprobes can help locate that area by recording the firing of neurons to a stimulus to a defined area during their placement in the spinal cord.

The depth to which microprobes are inserted depends on the fine structure of the area being examined. However, because they sample along their whole length one might as well insert them a generous distance.

193

Considerable care is needed in micromanipulator readings. It is not difficult to make an error of 100–200 μm when determining the manipulator reading for the surface of the spinal cord, particularly when the pia mater (and invariably some of the underlying neural tissue) has been locally removed. It also needs to be remembered that results from many microprobes will be averaged and this requires some constancy in the spatial relationships of structures which are possible sources of release. In practice this means placing microprobes in very nearly comparable sites in the nervous system. For example, in the spinal cord, the distance of the grey matter from the dorsal surface varies mediolaterally, and, hence, microprobes need to be inserted approximately at the same distance from the midline but with greater latitude in the rostrocaudal direction.

Dye should always be ejected at two sites at the conclusion of an experiment. Locating the resultant dye spots in histological sections can guard against some gross errors: for example missing the grey matter entirely!

4.2 Time in the central nervous system

This will need to be determined experimentally but some guides can be given:

- With antibodies to the carboxy terminus of substance P, pinching the skin of the hind paw gave clear evidence of spinal release with microprobes present in the spinal cord for 15 minutes (9). The same microprobes, however, detected release of substance P when only 5 minutes in the spinal cord and a chemically inflamed joint was subjected to pressure.

- Microprobes bearing antibodies to somatostatin showed spinal release within 5 minutes of pinching the skin.

- Microprobes bearing antibodies to neurokinin A (NKA) needed to be 30 minutes or longer in the spinal cord to detect stimulus-evoked release.

5. After the central nervous system

5.1 First wash

Following removal from the spinal cord each microprobe is clamped vertically in a holder and lowered into a continually stirred ice-cold solution of PBS–Tween solution and washed for 15 minutes. Each probe is then inserted into a 5 μl capillary containing a solution of a radiolabelled form of the peptide under study.

5.1.1 Labelled peptides

To date only ^{125}I-labelled peptides have been used with microprobes. This isotope has a half-life of 60 days and 10 μCi should be sufficient for 500–800 microprobes used over 6 weeks. When purchased already iodinated, the lyophylate is dissolved to a volume of 100 μl and 20 aliquots of 5 μl are stored frozen. Experience has shown that, to obtain X-ray film images of microprobes

Suction Block

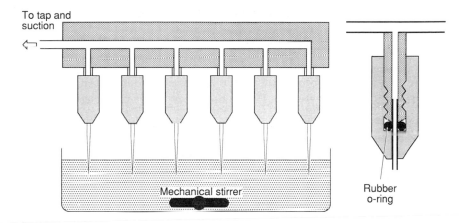

Figure 8. The final washing of six microprobes after incubation in radiolabelled neuropeptides. To the right is shown, in detail, the method of gripping each microprobe.

in a reasonable time (for example 3–5 days), the total radioactivity of the broken off tip of a microprobe bearing antibodies, and which has simply been incubated in a 5 µl capillary containing radiolabelled peptide, should be of the order of 1000 d.p.m. With the amounts of antibody commonly bound on the outer surfaces of microprobes, such radioactivity can be obtained by diluting each aliquot of labelled peptide to a volume giving 2000 d.p.m./µl. As an example, Bolton–Hunter [^{125}I]SP purchased with a specific activity of 2000 Ci/mmol would have approximately 500 µl of PBS containing 1% bovine serum albumin added to the first aliquot. The volume is progressively reduced as the sample ages. The level of radioactivity in 5 µl of sample of the diluted aliquot is checked in a gamma counter.

5.2 Final wash

Each microprobe with its attached capillary containing radiolabelled peptide is transferred to a tray in a cold room (as described previously under preparing microprobes) and incubated for 24 hours at 6 °C. Following this incubation, microprobes are washed, and the broken off tips are placed in an X-ray film cassette as described in *Protocol 5*.

The tips of most microprobes are blocked by tissue debris when introduced into the central nervous system, and, hence, little of the solution containing the labelled peptide enters the lumen of a microprobe during the 24 hour incubation. To guard against such a possibility, however, microprobes are washed with suction applied to the tips. Thus any contained labelled peptide entering through an unoccluded tip is massively diluted and sucked away.

Protocol 5. Washing microprobes

1. Fill a flat dish with Millipore-filtered PBS, containing Tween 0.1%, to a depth of 3 cm and stir magnetically.

2. Obtain 6 incubated microprobes from the cold room, remove the capillaries containing labelled peptide and dispose of them in an approved manner. The whole washing procedure should be done in a designated area and the normal precautions for handling radioactive materials observed.

3. Insert the open end of each microprobe through the rubber O-ring of a suction nozzle and tighten. *Figure 8* illustrates this in detail.

4. Lower the 6 mounted microprobes into the PBS–Tween, such that the terminal centimetre is covered, and wash with suction applied for 15 minutes. Microprobes with tips blocked will show no change in the appearance of contained dye. Those with patent tips will progressively pale from the tips.

5. After 15 minutes raise the suction block up and invert it while suction is still applied. This washes away liquids from within unblocked microprobes and replaces it with air.

6. Remove microprobes from the suction block.

7. Take a microprobe and gently break off (using both hands) approximately the terminal one centimetre. If this is done roughly the tip will shatter, which is disastrous since any zones of reduced binding cannot then be localized to a site in the nervous system. With experience there should be no microprobe losses at this stage.

8. Place the microprobe tip on a sheet of chromatography paper, cut to the size of the X-ray film being used, and attach it at its broken end with a small drop of typists' white-out solution. Too much adhesive could interfere with close apposition of X-ray film to a microprobe.

9. When all microprobes from an experiment are attached to the paper, place it in an X-ray film cassette (no intensifying screens) and, in a dark room with the appropriate safelight, add a sheet of monoemulsion X-ray film (for example Kodak NMC). Remember to locate the edge notch appropriately so that the emulsion side is in contact with the microprobes.

6. Analysis of microprobe images

6.1 Exposure times

Experience will decide how long to expose X-ray films to the microprobes. What is needed is an image well above the background silver grain density, but always below the maximal density for the film. In practice, using the materials and procedures described in this paper, satisfactory images will be obtained

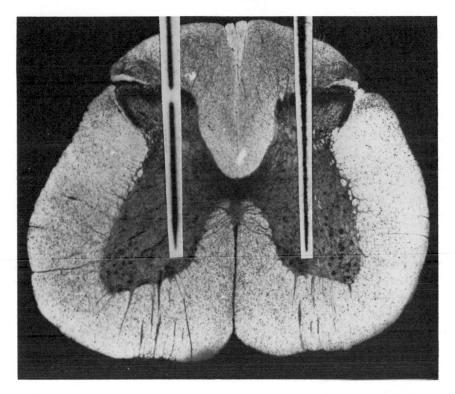

Figure 9. Enlargements of microprobe autoradiographs superimposed on a similarly enlarged print of the appropriate cross section of the spinal cord of the cat. These scattered images are greater in diameter than the actual microprobes. The microprobe on the right was inserted in the spinal cord and no peripheral stimulus was delivered. While the microprobe on the left was in the spinal cord a noxious mechanical stimulus was applied to the ipsilateral hind limb. Note the near complete inhibition of binding of [^{125}I] substance P in a discrete zone. Both microprobes bore antibodies to the carboxy terminus of substance P.

in 3–5 days. Occasionally one microprobe will have image scattering markedly greater than others, presumably resulting from inadequate contact with the film. In these cases gently scrape at the white-out fixative with a scalpel blade to free the relevant probe and reattach it at an adjacent site. The author usually obtains at least two films of each series of microprobes using different exposure times.

6.2 Simple photographic enlargements

Sheets of X-ray film bearing microprobe autoradiographs can be placed in an enlarger, the resultant prints can be superimposed on a print of a cross-section of the spinal cord to relate sites of release to particular structures. *Figure 9* illustrates such superimposed images. It should be noted that microprobes are much finer than the images shown on such reconstructions. The scattering associated with a radioactive source which is a finite distance above the film results in images appearing larger than sources. The difficulty here is attempting to give quantitative significance to inhibition of binding on autoradiographs. Such attempts depend on how images are printed and, hence, even on the contrast of the printing paper. The method can give useful results (1), but quantitative microdensitometric analysis of microprobe images is far superior.

6.3 Image analysis

Analysis of microprobe images is achieved with computer-assisted image analysis. In the author's laboratory the system was not bought prepackaged, but rather the components obtained separately and the programming done in-house. The principles are described in Hendry *et al.* (10).

Each microprobe is scanned by a charge-coupled device (CCD) camera using an enlarging lens held in an extendable bellows to give a large image. For microprobes inserted 3 or 4 mm into the spinal cord the terminal 5 mm occupies the whole width of a monitor screen. The camera and lens is held in a rack and pinion mount over the light source. The latter consists of a metal box with two 25 W microscope halogen bulbs on a movable platform within. Two white diffusion plates are located between the light bulbs and a microscope mechanical stage mounted on the top of the light source. Strips of film containing microprobe images, are placed on this mechanical stage. Only a light slit is allowed to illuminate a microprobe image. This is because a CCD camera will not operate as a satisfactory densitometer when scanning a microprobe image if the latter is only a small part of the total field. The slit consists of two movable metal plates positioned just below the mechanical stages. Two variable power supplies are used: one to set the gain of the CCD camera and the other to control the level of illumination.

The system employs an Imaging Technology PC Vision Plus board operating in an AT-based computer. Given the pace of change in image analysis, describing the system in detail would serve little purpose, but a concise description of the sequences used will enable a comparable (and probably faster) system to be assembled.

6.3.1 The code

Each microprobe image finishes up as an array of numbers, as shown in *Figure 10*. The first 32 come from a code which describes a host of experimental conditions relevant to each microprobe. These coded values are used to select defined microprobes from a file of images so that mean image analyses can be prepared. Coding sheets are used to prepare data for a number of microprobes before processing the images.

6.3.2 Image integration

The CCD camera used recognizes 256 degrees of brightness in an image, and, in the system used, does so in 10 μm-sided squares. By performing transverse integrations across a restricted part of the field which contains the microprobe

```
 52 101188   30    2    1   60    3    1    1    2    1   24    2    7    2    2    3   37   30   40   40   30
 20    2    2    1    0    8    5    3    2   99 1570 1733 1673 1786 1763 1873 1956 1915
1885 1879 1876 2088 2145 2085 2180 2184 2154 2099 2019 2115 2189
1994 1996 2131 2261 2280 2381 2518 2529 2583 2806 2869 2751 2676
2806 2756 2580 2571 2637 2681 2740 2749 3055 3103 2904 2876 2699
2620 2641 2763 2828 2784 2800 2903 2788 2757 2735 2640 2553 2557
2750 2636 2507 2593 2757 2790 2824 2803 2981 3008 3005 3002 2891
2938 2905 2746 2885 2817 2851 2980 2856 2971 2989 3089 3104 3105
3224 3243 3258 3406 3283 3200 3174 3082 3140 3207 3164 3093 2983
2903 2958 2949 3188 3206 3258 3299 3074 3056 2776 2679 2829 2845
3036 3086 3019 2993 2952 2891 2878 2893 2947 2989 2955 2919 2848
2767 2732 2806 2811 2763 2793 3034 2997 3002 3010 3018 2975 2823
2864 2790 2772 3021 3026 3094 3115 3047 2858 2771 2835 3071 3123
3233 3196 3189 3200 3105 2994 3027 3074 3269 3321 3375 3301 3180
3113 3071 3051 3211 3094 3250 3339 3243 3284 3300 3303 3451 3494
3551 3513 3377 3492 3307 3373 3368 3315 3321 3172 3184 3208 3156
3355 3287 3155 3214 3256 3314 3203 3110 3175 3241 3241 3219 3327
3413 3377 3366 3499 3459 3515 3622 3611 3616 3627 3523 3583 3504
3528 3590 3557 3590 3462 3399 3378 3484 3709 3454 3174 3220 3128
3083 3181 3328 3336 3040 3017 2949 2940 2879 2772 2711 2820 2742
2016 2844 2837 2857 2780 3038 3246 3165 3055 2855 2010 2678 2563
2315 2541 2546 2645 2602 2542 2501 2550 2564 2435 2284 2336 2311
2370 2358 2329 2388 2278 2268 2196 2315 2518 2405 2417 2435 2515
2634 2512 2604 2712 2575 2623 2505 2570 2498 2367 2379 2337 2240
2204 2108 2185 2130 2018 2033 2047 2086 2038 1870 1831 1658 1462
1502 1299 1387 1542 1436 1303 1197 1100 1099 1074 1313 1333 1250
1315 1264 1078  934  823  917  959 1019 1067 1073 1084 1137 1130 1124
1022  987  914  852 1021  990  924 1001  939  941  982  949 1098  919 1010
1104 1144 1290 1251 1241 1203 1139 1276 1302 1394 1351 1245 1178
1102 1125 1117 1092 1052 1007  933 1055 1015  931  970  876  828  871
 798  886  749  656  659  661  587  478  475  578  464  597  527  421  378  295
 318  427  481  595  517  456  608  564  808  809  621  773  636  648  690  719
 873  949  994 1127 1245 1457 1674 1733 1941 2203 2530 2935 3193
3516 3930 4109 4359 4406 4487 4633 4636 4632 4573 4656 4824 4851
5136 5252 5126 5086 5062 5135 5156 5286 5293 5015 4971 5023 4926
5128 5224 5171 5298 5184 5262 5318 5449 5561 5489 5405 5462 5420
5502 5334 5261 5250 5242 5305 5344 5375 5309 5093 5091 5170 5149
5193 5217 5338 5363 5220 5195 5195 5091 5063 4716 4555 4480 4367
4481 4670 4815 4836 4950 5062 5005 4885 4939 4918 4931 4945 5003
5153 5254 5253 5194 5112 5080 5061 5009 5139 5033 4952 5265    0
```

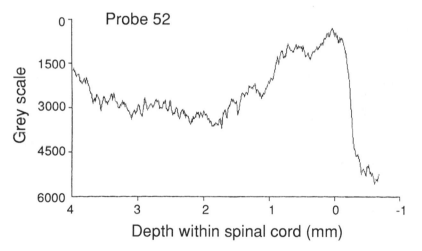

Figure 10. A file entry for a microprobe and the graph derived from this entry. The first 32 numbers encode data describing experimental conditions while the remainder describe the optical density of the microprobe autoradiograph and are plotted below. The numbers are plotted such that localized reduced binding of the labelled peptide produces an upward deflection.

image, an array of numbers is obtained which represents image blackness in 10 µm intervals.

Thus each file entry for a microprobe image has 532 numbers, 32 for experimental conditions and 500 for image integration (representing 5 mm of microprobe length). *Figure 10* shows one such entry.

The general sequence of events for image analysis is as follows:

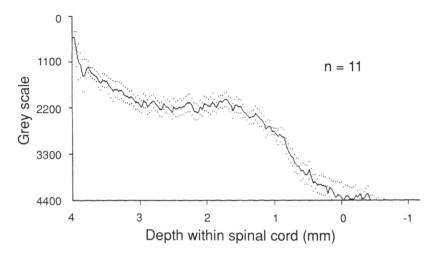

Figure 11. A mean image analysis of a group of selected microprobe images.

- Open a new or old file for data entry.
- Enter the coded values for a microprobe.
- Position the film containing the microprobe image over the light slit and illuminate a non-image (background) area.
- The camera and image board then scan the area selected and display the light intensity at a predetermined site. Adjust the light intensity to an empirically determined pre-set value. Average 10 background scans of one area and repeat this at another background area. Perform a median filter operation on the averaged backgrounds. A library of subroutines purchasable from the image board manufacturer enables this to be done.
- Place the microprobe image over the light slit with the tip in a fixed position and the orientation fixed, for example parallel to the upper edge of the monitor face, tip to the left, and within the upper one-fifth of the monitor field. Adjust the light intensity to the prefixed level and then average 10 scans.
- Perform background subtraction and then transverse integration.
- Display a graph of image density versus length, together with the coded values. If these are correct, enter into the file.

The data sorting program prepares mean image analyses. Thus one can select microprobes bearing, for example antibodies to the N-terminus of substance P, which were 15 minutes in the spinal cord and the peripheral stimulus was pinching a hind limb digit. Such a mean analysis is shown in *Figure 11*. For one reading of a file five such searches can be simultaneously compiled. Differences between control and experimental groups can be obtained, and statistical significance assigned to the differences at each point (again in 10 µm steps).

There are some limitations in the interpretation of mean image analyses. Averaging spatially requires a reference point, and with microprobes the reference is the tip. Hence, if there are differences in the spatial spread of anatomical laminae with respect to the tip this will result in a broadening of such a structure on a mean image analysis if it is a site of release. With large numbers of microprobes, however, the peak site of inhibition of binding should be correct. Although the image analysis is performed in 10 μm steps, the scattering inherent in an autoradiograph image, as a result of a source being a finite distance above the film, prevents such spatial resolution being attained. In practice, sites of release of 100 μm along a microprobe image should be discernable.

References

1. Duggan, A. W. and Hendry, I. A. (1986). *Neurosci. Lett.*, **68**, 134.
2. Duggan, A. W., Hendry, I. A. H., Green, J. L., Morton, C. R., and Hutchison, W. D. (1988). *J. Neurosci. Meth.*, **23**, 241.
3. Weetall, H. H. (1970). *Biochim. Biophys. Acta*, **212**, 1.
4. Wingard, I. B., Katchalski-Katzir, E., and Goldstein, L. (1976). *Applied biochemistry and bioengineering: Immobilized enzyme principles*, Vol. 1. Academic Press, New York.
5. Woodward, J. (1985). *Immobilized cells and enzymes—a practical approach*. IRL Press, Oxford.
6. Rochow, F. G. (1951). *An introduction to the chemistry of the silicones*. Wiley, New York.
7. Goding, J. W. (1978). *J. Immunol. Meth.*, **20**, 241.
8. Schaible, H-G., Jarrott, B., Hope, P. J., and Duggan, A. W. (1990). *Brain Res.*, **529**, 214.
9. Duggan, A. W., Morton, C. R., Zhao, Z. Q., and Hendry, I. A. (1987). *Brain Res.*, **403**, 345.
10. Hendry, I. A., Morton, C. R., and Duggan, A. W. (1988). *J. Neurosci. Meth.*, **23**, 249.

8

Metabolic mapping of
local neuronal activity

CHARLES KENNEDY, CAROLYN BEEBE SMITH,
and LOUIS SOKOLOFF

1. Introduction

The radioactive deoxyglucose method was developed to measure local rates of
energy metabolism simultaneously in all components of the brain of conscious
laboratory animals. It takes advantage of the extraordinary spatial resolution
made possible by quantitative autoradiography (1). The dependence on
autoradiography prescribed the use of radioactive substrates for energy
metabolism, the labelled products of which could be assayed in the tissues by
the autoradiographic technique. Although oxygen consumption is the most direct
measure of energy metabolism, the volatility of oxygen and its metabolic
products and the short physical half-life of its radioactive isotopes precluded
measurement of oxidative metabolism by the autoradiographic technique.

In most circumstances glucose is almost the sole substrate for cerebral
oxidative metabolism, and its utilization is stoichiometrically related to oxygen
consumption. Radioactive glucose is, however, not fully satisfactory because
its labelled products are lost too rapidly from the cerebral tissues. The labelled
analogue of glucose, 2-deoxy-D-[^{14}C]glucose (DG), was, therefore, selected
because its biochemical properties make it particularly appropriate to trace
glucose metabolism and to measure local cerebral glucose utilization by the
autoradiographic technique.

2. Theoretical basis of the deoxyglucose method

The method is based on the analysis of a model of the biochemical properties
of 2-deoxyglucose and glucose in brain (*Figure 1A*) (1). DG is transported bi-
directionally between blood and brain by the same carrier that transports glucose
across the blood–brain barrier. In the cerebral tissues it is phosphorylated by
hexokinase to 2-deoxyglucose-6-phosphate (DG-6-P). DG and glucose are,
therefore, competitive substrates for both blood–brain barrier transport and
hexokinase-catalysed phosphorylation. Unlike glucose-6-phosphate (G-6-P),

203

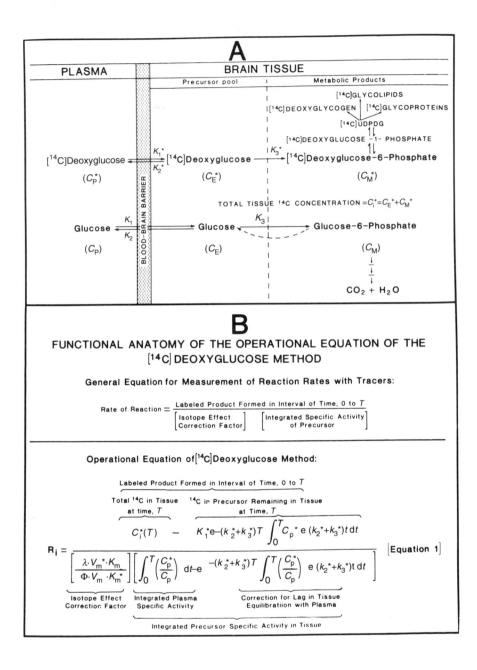

however, which is rapidly metabolized further to fructose-6-phosphate (F-6-P) and eventually to CO_2 and water, DG-6-P cannot be converted to F-6-P and is a poor substrate for G-6-P dehydrogenase. Furthermore, there is relatively little glucose-6-phosphatase (G-6-Pase) activity in brain and even less deoxyglucose-6-phosphatase activity that could hydrolyse DG-6-P back to free DG. DG-6-P can be converted into deoxyglucose-1-phosphate (DG-1-P), then into UDP-deoxyglucose (UDP-DG), and eventually into oligosaccharides, glycogen, glycolipids, and glycoproteins, but these reactions are slow, and in mammalian tissues only a very small fraction of the DG-6-P formed proceeds to these products (2). In any case, these compounds are all secondary products of DG-6-P; they are all relatively stable, and all of them together represent the accumulated products of DG phosphorylation. DG-6-P and its further products, once formed, remain, therefore, essentially trapped in the cerebral tissues, at least for the usual 45 minute duration of the experimental period.

If the experimental period is kept short enough, for example, less than one hour, to allow the assumption of negligible loss of labelled products from the tissues, then the quantity of ^{14}C-labelled products accumulated in any cerebral tissue at any given time following the introduction of $[^{14}C]DG$ into the circulation is equal to the amount of $[^{14}C]DG$ phosphorylated by hexokinase in that tissue during that interval of time. The amount of $[^{14}C]DG$ phosphorylated is in turn related to the amount of glucose phosphorylated over the same interval, depending on the time courses of the relative concentrations of $[^{14}C]DG$ and glucose in the precursor pools in the tissues and the Michaelis–Menten kinetic constants for hexokinase with respect to both $[^{14}C]DG$ and glucose. With glucose consumption in a steady-state, the amount of glucose

---

Figure 1. Theoretical basis of radioactive deoxyglucose method for measurement of local cerebral glucose utilization (1). A. Diagrammatic representation of the theoretical model. C_i^* represents the total ^{14}C concentration in a single homogeneous tissue of the brain. C_p^* and C_p represent the concentrations of $[^{14}C]$ deoxyglucose and glucose in the arterial plasma, respectively; C_E^* and C_E represent their respective concentrations in the tissue pools that serve as substrates for hexokinase. C_M^* represents the concentration of $[^{14}C]$deoxyglucose-6-phosphate in the tissue. The constants K_1^*, k_2^*, and k_3^* represent the rate constants for carrier-mediated transport of $[^{14}C]$deoxyglucose from plasma to tissue, for carrier-mediated transport back from tissue to plasma, and for phosphorylation by hexokinase, respectively; the constants K_1, k_2, and k_3 are the equivalent rate constants for glucose. $[^{14}C]$Deoxyglucose and glucose share and compete for the carrier that transports both between plasma and tissue and for hexokinase which phosphorylates them to their respective hexose-6-phosphates. The dashed arrow represents the possibility of glucose-6-phosphate hydrolysis by glucose-6-phosphatase activity, if any. B. Operational equation of the radioactive deoxyglucose method and its functional anatomy. T represents the time of termination of the experimental period; λ equals the ratio of the distribution space of deoxyglucose in the tissue to that of glucose; ϕ equals the fraction of glucose which once phosphorylated continues down the glycolytic pathway; and K_m^* and V_m^* and K_m and V_m represent the familiar Michaelis–Menten kinetic constants of hexokinase for deoxyglucose and glucose, respectively. The other symbols are the same as those defined in A.

phosphorylated during the interval of time equals the steady-state flux of glucose through the hexokinase-catalysed step times the duration of the interval, and the net rate of flux of glucose through this step equals the rate of glucose utilization.

These relationships can be rigorously combined into a model (*Figure 1A*) and mathematically analysed to derive an operational equation (*Figure 1B*), provided that the following assumptions are made:

- The tissue compartment is homogeneous with respect to the concentrations of [^{14}C]DG and glucose, blood flow, hexose transport, and glucose metabolism.

- The arterial plasma glucose concentration and rate of glucose consumption are constant throughout the experimental period.

- [^{14}C]DG concentrations in the plasma and tissues are at tracer levels (i.e. molecular concentrations of free [^{14}C]DG essentially equal to zero).

The operational equation which defines R_i, the rate of glucose utilization per unit mass of tissue i, in terms of measurable variables is presented in *Figure 1B*.

The rate constants, K_1^*, k_2^*, and k_3^*, in the operational equation (*Figure 1B*) are determined in a separate group of animals. This is done by fitting an equation (which defines the time course of tissue ^{14}C concentration in terms of time, history of arterial plasma [^{14}C]DG concentration, and these rate constants) to experimentally measured time courses of tissue and plasma concentrations of ^{14}C following administration of [^{14}C]DG (1). The λ, φ, and the Michaelis–Menten enzyme kinetic constants of hexokinase for DG and glucose are grouped together to constitute a single, lumped constant (*Figure 1B*). It can be shown, mathematically, that this lumped constant is equal to the asymptotic value of the product of the ratio of the cerebral extraction ratios of [^{14}C]DG and glucose and the ratio of the arterial blood to plasma specific activities when the arterial plasma [^{14}C]DG and glucose levels are maintained constant. The lumped constant is also determined in a separate group of animals from arterial and cerebral venous blood samples drawn during programmed intravenous infusions that produce and maintain constant arterial plasma [^{14}C]DG concentrations (1).

Despite its complex appearance, the operational equation is really nothing more than a general statement of the relationship used to determine rates of chemical reactions from measurements made with radioactive tracers (*Figure 1B*). The numerator of the equation represents the amount of radioactive product formed in a given interval of time; it is equal to $C_i^*(T)$, the combined concentrations of [^{14}C]DG and [^{14}C]DG-6-P and subsequent labelled products in the tissue at time T, measured by the quantitative autoradiographic technique, minus a term that represents the free unmetabolized [^{14}C]DG still remaining in the tissue. The denominator represents the integrated specific activity of the precursor pool times a factor, the lumped constant, which is analogous to a correction factor for an isotope effect; it corrects for the differences in the kinetics of blood–brain barrier transport and phosphorylation between DG and glucose. The term with the exponential factor in the denominator takes into account the lag in the equilibration of the tissue precursor pool with the plasma.

3. Procedure

The operational equation (*Figure 1B*) dictates the variables that need to be measured to determine the local rates of cerebral glucose utilization. The specific procedure employed is designed to evaluate these variables and to minimize potential errors that might occur in the actual application of the method. If the rate constants, K_1^*, k_2^*, and k_3^*, are precisely known, then the equation is generally applicable with any mode of administration of [^{14}C]DG and for a wide range of time intervals. The values of the rate constants determined in different groups of conscious rats with different arterial plasma glucose levels varying between normoglycaemia and hyperglycaemia are presented in Table 1. The values of the rate constants are comparable in the monkey (3), dog (4), sheep (5), and man (6). These rate constants vary with blood flow and transport and the rate of phosphorylation of DG, and can, therefore, be expected to vary with the condition of the animal; for most accurate results they should, therefore, be redetermined for each condition studied. Determination of the rate constants requires, however, a rather extensive experimental procedure involving numerous animals. The structure of the operational equation suggests a more practical alternative to the determination of the rate constants for each condition. All the terms in the equation that contain the rate constants approach zero with increasing time if the [^{14}C]DG is so administered that the plasma [^{14}C]DG concentration also approaches zero with time. From the values of the rate constants determined in normal animals and the usual time course of the clearance of [^{14}C]DG from the arterial plasma following a single intravenous pulse at zero time, it has been determined that an interval of 30–45 minutes after a pulse is adequate for these terms to become sufficiently small that considerable latitude in inaccuracies of the rate constants is permissible without appreciable error in the estimates of local glucose consumption (1). An additional advantage of the use of a pulse of [^{14}C]DG followed by a relatively long interval before killing the animal is that by then most of the free [^{14}C]DG in the tissues has been either converted to [^{14}C]DG-6-P or transported back to the plasma; the optical densities in the autoradiograms then represent mainly the concentrations of [^{14}C]DG-6-P and are, therefore, images of the relative rates of glucose utilization in the various cerebral tissues.

Table 1. Values of rate constants for normoglycaemic and hyperglycaemic rats[a]

Rate constants	Arterial plasma glucose level (Mean ± SD)			
	8 ± 2 mM[c]	20 ± 1 mM[d]	25 ± 1 mM[d]	31 ± 1 mM[d]
Grey matter:				
K_1^* (ml/g/min)	0.189 ± 0.012	0.127 ± 0.006	0.116 ± 0.005	0.085 ± 0.002
k_2^* (min^{-1})	0.245 ± 0.040	0.235 ± 0.019	0.240 ± 0.022	0.231 ± 0.015
k_3^* (min^{-1})	0.052 ± 0.010	0.042 ± 0.004	0.026 ± 0.003	0.027 ± 0.002
$\log_e 2/(k_2^* + k_3^*)$ (min)[b]	2.4	2.5	2.6	2.7
White matter:				
K_1^* (ml/g/min)	0.079 ± 0.008	0.057 ± 0.005	0.052 ± 0.004	0.041 ± 0.002
k_2^* (min^{-1})	0.133 ± 0.046	0.168 ± 0.034	0.120 ± 0.018	0.148 ± 0.023
k_3^* (min^{-1})	0.020 ± 0.020	0.030 ± 0.008	0.004 ± 0.006	0.013 ± 0.006
$\log_e 2/(k_2^* + k_3^*)$ (min)[b]	4.5	3.5	5.6	4.3

[a] The values for the rate constants represent the means ± SEM of the values for each rate constant obtained in 15 grey structures and 3 white structures in the normoglycaemic (7–12 mM) group of 15 animals and in 16 grey and 2 white structures of the other groups of 30 animals each. The rate constants for each structure were obtained by non-linear least squares fitting to the data obtained from all the animals for each group. The values for the SEM were calculated from the standard errors of the estimates of the individual rate constants and their covariances.
[b] $\log_e 2/(k_2^* + k_3^*)$ represents half-life of free [^{14}C]DG in the tissue.
[c] From Sokoloff *et al.* (1).
[d] From Orzi *et al.* (20).

Protocol 1. Preparation of animals and procedure of deoxyglucose method

Equipment and supplies for measuring local cerebral glucose utilization

- apparatus for administering anaesthetic, e.g. halothane
- surgical instruments and plastic catheters of suitable size (e.g. PE-50) for catheterization of artery and vein
- animal restraining system (see text)
- stop watches
- small centrifuge and heparinized, fluoride-treated centrifuge tubes (e.g. Beckman Microfuge B or equivalent)
- mercury or electronic manometer to measure arterial blood pressure
- capillary kits for sampling blood for measurement of arterial blood hematocrit, pH, PCO_2, and PO_2 (e.g. Ciba–Corning Diagnostics)
- crushed ice, liquid nitrogen or dry ice, and isopentane
- pipettes (10 and 20 µl) for measuring plasma samples and equipment for measuring plasma glucose and ^{14}C concentrations

- equipment for sectioning frozen brain (e.g. cryostat, microtome, coverslips, hot plate, etc.)
- calibrated [^{14}C]methylmethacrylate standards for quantitative autoradiography (see *Protocol 2*).
- equipment for quantitative autoradiography, e.g. dark room, X-ray film (e.g. Kodak SB-5, OM-1, etc.), and cassettes, etc., as described in text
- densitometer or computer-assisted image processing system
- computer for calculations of local cerebral glucose utilization

Steps in procedure:
1. Catheterize an artery and vein under light halothane anaesthesia. Allow at least 3 hours for recovery (Section 3.1).
2. Inject intravenous pulse of [^{14}C]DG (no more than 2.5 µmol/kg) dissolved in physiological saline solution, and collect timed samples of arterial blood continuously in as many discrete samples as possible from zero time to 45 seconds (to catch the peak) and then at 1, 2, 3, 5, 7.5, 10, 15, 20, 25, 30, 35, and 45 minutes (Section 3.2).
3. Centrifuge samples to separate plasma and store on ice until analysed.
4. Kill animal at approximately 45 min by decapitation or by intravenous thiopentone, followed immediately by a saturated solution of KCl.
5. Remove and freeze brain in isopentane at $-40\,°C$ and store at $-70\,°C$.
6. Analyse samples of arterial plasma for their concentrations of [^{14}C]deoxyglucose and glucose (Section 3.3).
7. Cut brain into 20 µm sections in cryostat at -20 to $-25\,°C$, thaw-mount on glass coverslips, and dry sections on hot plate at $60\,°C$ (Section 3.4).
8. Expose sections and calibrated [^{14}C]methylmethacrylate standards to X-ray or photographic film in suitable cassettes. Develop film (Section 3.5).
9. Analyse autoradiograms by quantitative densitometry and, together with plasma data and known constants, calculate local glucose utilization by the operational equation of the method (Section 4).

3.1 Preparation of animals

Under general anaesthesia insert catheters into any conveniently located artery and vein (*Protocol 1*). Halothane is the preferred anaesthetic because of the relatively short recovery period and lack of prolonged metabolic after-effects. The catheters must have properties that allow them to remain patent for repeated blood sampling. Those made of polyethylene in the size designated as PE-50 by the supplier, Clay Adams Division of Becton, Dickinson and Company, are quite satisfactory for monkeys, cats, dogs, and adult rats; with smaller animals the narrower sized PE-10 may be necessary. The catheters are filled with bubble-free dilute heparin solution (100 units/ml in normal saline) and plugged at the distal end before insertion; these generally remain patent for several hours. To minimize blood loss when flushing the dead space of the catheter system before withdrawal of each sample, it is desirable to keep the arterial catheter as short as possible, for example 15–18 cm or less for the adult rat. After insertion of the catheters, suture the wounds, and apply 5% lidocaine ointment locally. During recovery from anaesthesia the animal may be placed in a suitable restraining device. In the case of rats, a loosely fitting, bivalved, plaster cast around the lower trunk is applied with the hind legs taped to a lead brick. For cats, a zippered jacket is satisfactory; for monkeys, a restraining chair is used. For behavioural studies that require a freely-moving animal the catheters may be threaded under the skin to exit at the nape of the neck (7, 8), and the animals remain unrestrained. If attention is paid to keeping the concentration of halothane to a minimum, and the time for the surgical procedure to 15–20 min, recovery is rapid, and the experiment can be initiated within 2–3 h. Immediately before starting the experiment it is advisable to measure body temperature, mean arterial blood pressure, and arterial hematocrit, pH, PCO_2, and PO_2 to evaluate the physiologic state of the animal.

3.2 Administration of [^{14}C]deoxyglucose and sampling of arterial blood

To ensure that tracer conditions are maintained, the dose of [^{14}C]DG should be limited to no more than 2.5 µmol of DG per kg of body-weight. If the specific activity is relatively high (for example 50–60 mCi/mmol), 100–125 µCi per kg of body-weight can be given; with normal rats, for example, this amount of isotope is sufficient to attain desirable optical densities in the autoradiograms within a reasonable time, i.e. 4–6 days of exposure with Kodak SB-5 X-ray film. Economic factors may dictate the use of lower doses, but at the expense of longer exposure times. The [^{14}C]DG is usually supplied in an ethanol solution; it must first be evaporated to dryness and the DG then redissolved in physiological saline before injection. A suitable concentration for its intravenous injection is 100–200 µCi/ml.

The experimental period is initiated by the infusion of [^{14}C]DG through the venous catheter over a period of 10–15 s. With zero time marking the start of the infusion, begin sampling of arterial blood from the arterial catheter immediately, and collect blood continuously during the first 45 s in consecutive samples into heparinized, fluoride-treated plastic centrifuge tubes to capture the peak. Discrete arterial samples are then subsequently collected at the times specified in *Protocol 1*. The volume of each blood sample should be sufficient for assays of [^{14}C]DG and glucose concentrations in the plasma (for example 50–100 µl). Because the plasma glucose concentration usually changes little during the experimental period, it may be sufficient to determine plasma glucose concentration in samples taken at four or five time points (for example 0, 10, 25, 35, and 45 min) and to interpolate from these the values for the intermediate times. By so doing it is possible to minimize the amount of blood taken from the animal. Care must be taken to clear the dead space of the catheter prior to each sample. The blood samples are immediately centrifuged in a small, high-speed centrifuge such as the Beckman Microfuge B to separate the plasma, they are then kept on ice until the plasma is analysed. The use of heparinized centrifuge tubes may be unnecessary if the animal is heparinized just prior to the experiment.

Blood loss during the blood sampling and clearing of the dead space should be minimized because shock is readily induced in small animals. Most rats weighing 300–400 g can tolerate the removal of 2 ml over the 45 min period without development of significant hypotension. Blood drawn for the purpose of clearing the dead space (for example three times the volume of the dead space in the catheter) may, of course, be returned to the animal. Because of the occasional rat that fails to tolerate even small blood losses, it is well to monitor mean arterial pressure at intervals during the sampling period. A fall below 90 mmHg is reason to eliminate the animal from a study unless the effects of hypotension are being examined.

Immediately after the last blood sample is drawn, kill the animal either by decapitation in the case of small animals or, alternatively, by an intravenous injection of thiopentone followed immediately by a saturated solution of KCl to stop the heart. Immediately remove the brain and freeze in a bath of isopentane (2-methylbutane) prechilled to − 40 to − 50 °C with dry ice or liquid nitrogen. Normal external configuration of the brain can be maintained during the freezing process if small brains are lowered into the isopentane with forceps holding the cervical cord, and large brains are placed in a ladle shaped approximately to the brain's contour and coated with a film of mineral oil to facilitate removal after freezing.

3.3 Analysis of arterial plasma for [^{14}C] deoxyglucose and glucose concentrations

Plasma [^{14}C] DG concentration is measured by liquid scintillation counting of the ^{14}C content. Twenty microlitres of plasma are pipetted into 1 ml of water contained in a counting vial. Ten millilitres of a suitable phosphor solution are added (for example Aquasol, Du Pont). With internal or external standardization of the counting efficiency the d.p.m. are determined and the concentration then expressed in nCi/ml of plasma. The plasma glucose concentration is most conveniently assayed in a glucose analyser (for example Glucose Analyzer 2, Beckman Instruments). This is quick and requires only 10 µl of plasma per determination. Also suitable for this purpose is the coupled, glucose-dependent, hexokinase–glucose-6-phosphate dehydrogenase-catalysed reduction of NADP$^+$ which is available in kit form from Calbiochem.

3.4 Processing the brain tissue

Some investigators have chosen to perfuse the brain with a fixative immediately after the animal is killed. This serves to improve the quality of the histologic sections which facilitates anatomical identification of regions of interest in the autoradiograms. Perfusion fixation probably also reduces artefacts in the sectioning of the frozen brain, especially those of large animals. For this purpose 3.5% formaldehyde in 0.05 M phosphate buffer, pH 7.4, may be employed. The animal is heparinized immediately before killing. With a cannula placed in the left ventricle and the reservoir of the perfusate about 90–100 cm (i.e. equivalent to pressures of about 65–75 mmHg) above the heart level, the perfusion is carried out for 10 min. The brain is then removed and frozen as described below.

We have attempted to determine whether perfusion fixation alters the distribution of ^{14}C in the brain; autoradiograms made from sections of non-perfused brain showed clearer definition of fine cortical markings, and the perfusion appeared to wash out as much as 15% of the label. Until a perfusion procedure is devised which prevents loss or movement of the label in the tissue, it is recommended that perfusion be omitted and the brain be frozen immediately after removal.

After freezing, small brains are mounted on microtome tissue holders with an embedding matrix, such as that supplied by Lipshaw Mfg Co. Brains of larger animals must be cut into two or three smaller blocks in order for them to be accommodated in the microtome. Cutting frozen brain into blocks is best done with a small band saw, the blade of which is prechilled immediately before the cut is made. It is crucial to maintain the brain below $-30\,°C$ at all times during handling to prevent movement of the label by diffusion. Alternatively, blocking can be carried out before freezing. The head with the dorsal aspect of the skull open can be mounted in a stereotaxic device and the brain cut with a large microtome blade guided by needles inserted into the tisue stereotaxically. For prolonged storage prior to sectioning the brain should be kept in a sealed plastic bag in a freezer maintained at $-70\,°C$. The brain is sectioned at a thickness of 20 µm with a microtome in a cryostat maintained at a temperature of about -20 to $-22\,°C$. Even at this low temperature there is some movement of the label in a matter of hours, dictating the need to complete the sectioning of one block or an entire small brain, as the case may be, at one sitting. Errors due to inconsistent thickness of sections as well as a number of other artefacts can be introduced at this stage unless attention is given to a number of operational details of the microtome; these include secure mounting, knife sharpness, anti-roll plate adjustment, and the use of smooth, regularly timed strokes of the knife. The 20 µm sections are picked up from the knife surface on glass coverslips and thaw-mounted. They are then immediately transferred to a hot plate maintained at $60\,°C$ on which the thawed sections become dry within 5–8 s. The coverslips are then placed on an adhesive-coated paper board that has been cut to fit in a $10'' \times 12''$ X-ray cassette with a centre gap left for a strip holding a set of previously calibrated [^{14}C] methylmethacrylate standards (see *Protocol 2*) that span the range of ^{14}C concentrations in the tissues. The thickness of the backing which holds the standards must be adjusted so that contact with the emulsion of all surfaces containing radioactivity will be uniform.

An alternative system of sectioning brain is that which employs an LKB 2250 PMV cryomicrotome (Pharmacia LKB, Piscataway, USA). This has a number of advantages, especially in processing brains of large animals. Because of its capability of cutting bone the need for removal of the brain from the calvarium is eliminated. The entire head is sectioned, and, thus, the brain's normal relationship to other tissues is preserved. It has the additional advantage of eliminating various cutting artefacts, especially those resulting from wrinkling or fragmentation. Sections are remarkably uniform in thickness. Other artefacts, however, may be introduced, the most troublesome of which is that caused by slight shrinkage of the tissue on drying which results in fine lines giving the appearance of parched mud. Some users of this system have noted considerable loss of resolution of detail in autoradiograms. The reasons for this are uncertain, but it is possible that it results from transient warming of the brain at the time the head is mounted in the embedding medium or from the prolonged period necessary for completion of sectioning. Further experience with the system will determine whether or not this serious drawback can be overcome.

3.5 Preparation of autoradiograms

The 20 µm thick dried brain sections are autoradiographed in the cassettes on photographic film together with a set of calibrated [^{14}C] methylmethacrylate standards that span the range of tissue ^{14}C concentrations anticipated in the tissues. The procedure for calibrating the standards is described below and in *Protocol 2*. A variety of film types are suitable for contact autoradiography with ^{14}C. The single-coated, blue sensitive SB-5 X-ray film made by Eastman Kodak is generally satisfactory when developed according to the manufacturer's instructions. If the dose and specific activity of [^{14}C] DG suggested above are employed, satisfactory images having an optical density range between 0.1 and 1.0 are generated in 4–6 days. Films with finer grain, but requiring longer periods of exposure, are Eastman Kodak's OM-1, EMC-1, and Plus-X.

3.6 Calibration of autoradiographic standards

Quantitative autoradiography requires the simultaneous exposure of calibrated standards with the brain sections on each film (*Figure 2*). Small squares cut from sheets of [^{14}C] methylmethacrylate having a uniform distribution of ^{14}C are available from several commercial sources. It is important to note that their designated ^{14}C concentrations in µCi per g of plastic material supplied by the manufacturer are not the calibration values that should be used in the determinations of tissue ^{14}C concentrations from the optical densities in the

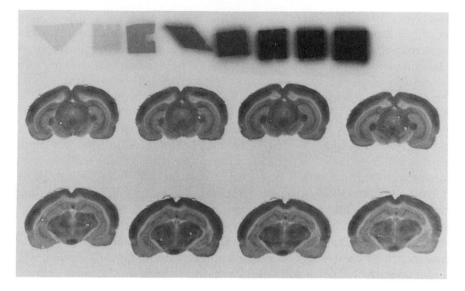

Figure 2. Autoradiograms of [^{14}C] deoxyglucose labelled sections of conscious rat brain and the calibrated [^{14}C] methylmethacrylate standards used to quantify ^{14}C concentration in tissues. It is sometimes difficult to avoid slight variation in section thickness which accounts for some variation in optical density from section to section. Where this is present, measurements of optical density must be made in several sections for any given structure.

autoradiograms. Each standard must be assigned a calibration value that is equivalent to that concentration of ^{14}C in whole brain, a 20 µm thick dried section of which would produce the same optical density as the standard when autoradiographed together. The standards must first be calibrated against brain tissues with known ^{14}C concentrations. This is done by autoradiographing the standards on the same films together with brain sections prepared, as described above, from brains in which the ^{14}C concentration has been directly measured. To be sure that the concentrations of ^{14}C are the same in the portions of brain tissue autoradiographed and in the portions directly assayed, it is necessary to use a ^{14}C-labelled compound and a procedure that labels the brain uniformly. [^{14}C] Antipyrine, [^{14}C] iodoantipyrine, and 3-O-[^{14}C] methylglucose have been found to be satisfactory if sufficient time is allowed for equilibration between the brain tissue and the blood. At equilibrium the concentration of all of these compounds is essentially uniform throughout the brain. A dose of any of these, given intravenously as a pulse, equilibrates in the brain within 60–120 min. It may be more complex, but it is preferable to administer the tracer by a programmed intravenous infusion that produces and maintains a constant arterial concentration (9); full and more stable equilibration is then achieved much sooner, i.e. less than 60 minutes. After equilibration the animal is killed and the brain removed. One hemisphere is immediately weighed and homogenized in a 1:30 (v/v) solution of Triton X-100 in water. The suspension

214

is then diluted with the Triton X-100 solution in a volumetric flask to a specific volume (for example 25 ml/g brain tissue). The suspension of homogenized brain tissue is then assayed for its ^{14}C concentration in a liquid scintillation counter with internal or external standardization. With a correction for the dilution, the ^{14}C concentration expressed in nCi/g of wet weight of brain is obtained; this value is the reference against which the plastic standards will be calibrated. The other hemisphere of each brain is frozen, sectioned at 20 μm, dried, and autoradiographed as described above. Brain sections so prepared from a number of animals given various doses of the tracer (for example 10–1000 μCi per 300–400 g rat) to provide a wide range of tissue concentrations and the set of methylmethacrylate standards to be calibrated are then autoradiographed together on the same film. The optical densities for the brain sections and the plastic standards are then measured by densitometry. The brain sections, for which the whole brain ^{14}C concentrations were directly measured by liquid scintillation counting of the opposite hemispheres, are then the primary standards against which the plastic standards are calibrated. The optical densities of the brain sections are, therefore, plotted against the measured concentrations of the brain from which they were cut. A smooth curve that is closely fitted by a cubic equation is generally obtained. The brain equivalent calibration values for the plastic standards are then obtained from this curve on the basis of their optical densities.

Protocol 2. Calibration of autoradiographic standards

1. Catheterize artery and vein in a group of rats (300–400 g) under halothane anaesthesia as in *Protocol 1* and allow 1–2 hours for recovery.

2. Inject ^{14}C-labelled tracer compound (e.g. [^{14}C]antipyrine, [^{14}C]iodoantipyrine, or 3-O-[^{14}C]methylglucose) intravenously as described in text; vary the dose from 10–1000 μCi among the various rats.

3. Wait 1–2 h for blood–brain equilibration, and then kill the rat and remove the brain.

4. Divide forebrain into two hemispheres.
   - Freeze one hemisphere in isopentane at −40 to −50 °C.
   - Weigh other hemisphere immediately to avoid evaporation and thoroughly homogenize in 5 ml of 1:30 (v/v) Triton X-100 in water.
   - Transfer total homogenate to 25 ml volumetric flask and dilute to mark with Triton X-100 solution.
   - Carefully transfer measured 20–50 μl samples of diluted brain suspension to counting vials and count in liquid scintillation counter with internal or external standardization.
   - Calculate concentration of ^{14}C in nCi/g of wet brain for each brain.

5. Cut 20 μm thick sections fom the frozen hemispheres of all the brains, dry, and autoradiograph (as described in *Protocol 1* and text) together with the set of [^{14}C]methylmethacrylate standards to be calibrated.

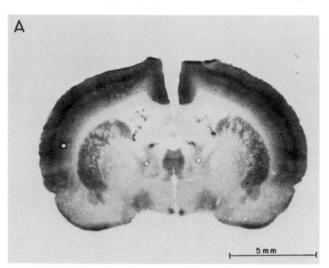

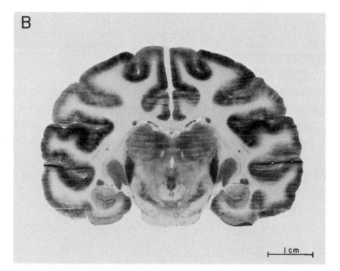

Figure 3. Autoradiograms of coronal sections of conscious rat (A) and monkey (B) brain made with [^{14}C]deoxyglucose prepared as described in the text. The pattern produced by differences in optical density permits identification of most of the component structures. The two small dark spots in the centre at the base of the brain in A correspond to the suprachiasmatic nuclei. Layers in the cerebral cortex of the rat can also be distinguished, as well as radial markings perpendicular to the cortical surface. In both autoradiograms the right side of the brain is on the right side of the figure. Note the right–left symmetry in optical density in all regions except for a region of cortex in the monkey (B). The monkey was engaged in a task involving the left arm throughout the experiment resulting in increased glucose utilization in the primary motor and sensory regions of the cortex.

216

Protocol 2. *Continued*

6. Measure optical densities of autoradiograms of [^{14}C]methylmethacrylate standards and brain sections; plot optical densities of brain sections against concentrations of ^{14}C measured in other hemispheres of the brains.

7. Determine equivalent brain tissue concentrations for plastic standards from calibration curve on the basis of optical densities of the plastic standards.

---

4. Analysis of autoradiograms and calculation of rates of glucose utilization

The autoradiograms of the brain sections provide pictorial representations of the relative rates of glucose utilization in the various structures of the brain, the darker the region the higher the rate of glucose utilization (*Figures 2 and 3*). Verification of the anatomical identity of a region can be made by histologic examination of the section from which the autoradiogram was made. With attention paid to the details cited above, the technique delineates relatively small structures of the rat brain, such as the suprachiasmatic nucleus (about 300 μm in coronal section) and the medial and lateral habenulae. In the normal rat cerebral cortical Layer IV, which measures about 100–200 μm, is seen as a dark linear band (*Figure 2*). Measurements have also been made in individual layers of Ammon's horn (10).

The determination of the rate of glucose utilization for any given region of the autoradiogram requires the densitometric determination of the concentration of ^{14}C in that region. Transmission densitometers of the type widely used in photography are suitable for this purpose; apertures of 0.2–0.5 mm in diameter are desirable. For satisfactory readings to be made in very small structures a computerized image processing system is necessary (see below). Whatever system is employed for densitometry, it is necessary to make several optical density readings for a given structure in several sections. The mean value of these readings serves both to reduce errors due to variations in section thickness and to give a value which is reasonably representative of the structure in its three dimensions. The concentration of ^{14}C in each structure is determined by comparison of its optical density with a calibration curve obtained by plotting the optical density of the calibrated plastic standards against their equivalent brain ^{14}C concentrations. Each film, of course, has its own standard calibration curve.

The operational equation for the calculation of glucose utilization is given in *Figure 1B*. The measured variables are:

● The entire time course of the arterial plasma [^{14}C]DG concentration, C_P^*, for zero time to the time of killing, T.

Table 2. Values of lumped constant in various species

Animal	No. of animals	Mean $\pm$ SEM
Albino rat:		
normal conscious	15	0.464 ± 0.026^a
anaesthetized	9	0.512 ± 0.039^a
conscious (5% CO_2)	2	0.463 ± 0.086^a
combined	26	0.481 ± 0.023
Monkey (conscious)	7	0.344 ± 0.036
Cat (anaesthetized)	6	0.411 ± 0.005
Dog (Beagle puppy) (conscious)	7	0.558 ± 0.031
Sheep:		
fetus	5	0.416 ± 0.014^b
newborn	4	0.382 ± 0.012^b
combined	9	0.400 ± 0.011
Man (conscious)	6	0.568 ± 0.043

[a] No statistically significant differences between normal conscious and anaesthetized rats ($0.3 < p < 0.4$) and between normal conscious rats and those breathing 5% CO_2 ($p > 0.9$).
[b] No statistically significant difference between fetal and newborn sheep ($p > 0.05$).

Note: Values obtained from the literature. For individual references see Sokoloff (60).

- The steady-state arterial plasma glucose level, C_P, over the same interval.
- The concentration of ^{14}C in the tissue at the time of killing, $C_i^*(T)$, which is determined by densitometric analysis of the autoradiograms.

The rate constants, K_1^*, k_2^*, and k_3^*, and the lumped constant are not measured in each experiment; the values for these constants are those which have already been determined and are characteristic of the species of animal. For the reasons given below the average rate constants for all grey or white structures for the condition studied may be used (*Table 1*). The values for the lumped constant to be used for various species are presented in *Table 2*. A programmable calculator or computer is employed to calculate the values for glucose utilization from the operational equation.

5. Computerized image processing

The autoradiograms contain an immense amount of information that cannot be practically recovered by manual densitometry or adequately represented by tabular presentation of the data. A computerized image-processing system may be used to analyse and transform the autoradiograms into colour-coded pictorial maps of the rates of local glucose utilization throughout the CNS (11). The autoradiograms are scanned automatically by a computer-controlled scanning microdensitometer. The optical density of each spot on the autoradiogram, from

25 to 100 µm as selected, is stored in a computer, converted to ^{14}C concentration on the basis of the optical densities of the calibrated ^{14}C plastic standards, and then converted to local rate of glucose utilization by solution of the operational equation. Colours are assigned to narrow ranges of the rates of glucose utilization, and the autoradiograms are then displayed on a monitor in pseudo-colour along with a calibrated colour scale for identifying the rate of glucose utilization in each spot of the autoradiogram from its colour. The images display the local rates of glucose utilization in all parts of the brain encoded in colour and delineate virtually all macroscopic neuroanatomical subdivisions of the brain. In recent years a variety of image-processing devices have become available.

6. Interpretation of autoradiograms

Certain considerations must be taken into account in interpreting the auto-radiograms and the calculated glucose utilization rates for the various regions of brain. Although it has been clearly demonstrated that changes in neuronal activity (for example by experimental stimulation or induced behaviour) result in changes in the local rates of glucose utilization (12), the various rates found in different regions of brain of inactive, unstimulated animals cannot be interpreted necessarily to reflect variations in neuronal or functional activity. As illustrated in deoxyglucose studies in which local functional activities have been markedly depressed, for example deafferented primary visual cortex (13), deep anaesthesia (1), and brain slices *in vitro* (14), there is a basal rate of glucose utilization which varies from one region to another. Such variation is independent of function and appears to reflect the tissues' energy requirements for maintenance of structure and biosynthetic processes. Superimposed upon such a basal level of energy metabolism is an increment associated with functional activity, conscious awareness, and behaviour. Thus, the distribution of rates in such awake, conscious animals cannot by itself be considered to provide maps of local functional activity. It is only when the results obtained in animals in which specific neural systems have been stimulated or depressed are compared with those in control animals that functional activity is mapped. The experimentally-induced changes in local glucose utilization are essentially evoked metabolic responses, and when localized, they map local functional activities.

Exact knowledge of the cellular elements participating in evoked metabolic responses must await technical improvements in the spatial resolution of the autoradiograms. To date little information has been provided to determine to what extent glia, for instance, are involved, but there is considerable evidence that induced changes in metabolism take place primarily in neuropil rather than in cell bodies (15–17). The comparatively low metabolic rates found in myelinated pathways have led some to attribute to white matter a fixed, resting rate. This is clearly not the case, and, indeed, the relative response to induced

Table 3. Local rates of cerebral glucose utilization in the normal, conscious albino rat and monkey

Structure	Albino rat (10)[a]	Monkey (7)[b]
Grey matter		
Visual cortex	107±6	59±3
Auditory cortex	162±5	79±4
Parietal cortex	112±5	48±4
Sensory–motor cortex	120±5	44±3
Thalamus: lateral nucleus	116±5	54±2
Thalamus: ventral nucleus	109±5	43±2
Medial geniculate body	131±5	65±3
Lateral geniculate body	96±5	39±1
Hypothalamus	54±2	25±1
Mammillary body	121±5	57±3
Hippocampus	79±3	39±2
Amygdala	52±2	25±2
Caudate–putamen	110±4	52±3
Nucleus accumbens	82±3	36±2
Globus pallidus	58±2	26±2
Substantia nigra	58±3	29±2
Vestibular nucleus	128±5	66±3
Cochlear nucleus	113±7	51±3
Superior olivary nucleus	133±7	63±4
Inferior colliculus	197±10	103±6
Superior colliculus	95±5	55±4
Pontine grey matter	62±3	28±1
Cerebellar cortex	57±2	31±2
Cerebellar nucleus	100±4	45±2
White matter		
Corpus callosum	40±2	11±1
Internal capsule	33±2	13±1
Cerebellar white matter	37±2	12±1
Weighted average for whole brain		
	68±3	36±1

Note: The values are means ± SEM (in µmole/100 g/min) of values determined in number of animals indicated in parentheses.

[a]From Sokoloff *et al.* (1).
[b]From Kennedy *et al.* (3).

stimulation in white matter may be of a magnitude similar to that in grey matter.

7. Normal rates of local cerebral glucose utilization in conscious animals

Local rates of glucose utilization in brain have been measured in the normal conscious and anaesthetized albino rat (1), in the conscious Rhesus monkey (3), fetal and neonatal sheep (5), and the newborn dog (4). The distribution of these rates in the various regions of brain is similar in all animals. Values in white matter are distributed in a narrow low range, and those in grey structures are broadly distributed around an average about 3–4 times greater than that of white matter (*Table 3*). The highest rates are found in the structures of the auditory system, with the inferior colliculus clearly the most metabolically active. Values for homologous regions on the two sides are normally virtually identical, and failure to find such symmetry in normal animals suggests the presence of disease or defective development, such as may be found in the relay nuclei of the auditory pathway in rats with unilateral otitis media, a common disorder in rats.

The rates of local cerebral glucose utilization in the conscious monkey exhibit similar heterogeneity to those found in the rat, but they are generally one-third to one-half the values in corresponding structures (*Table 3*). The differences in rates in the rat and monkey brain are consistent with the different cellular packing densities in the brains of these two species. Concern has been expressed in some quarters that restraint may result in rates of cerebral glucose utilization which are above those of unrestrained animals because of 'stress'. When, however, the values for local cerebral glucose utilization in freely-moving rats and in rats in the plaster-cast restraining system described above are compared, the results are almost identical (8, 18).

8. Theoretical and practical considerations

The design of the DG method was based on an operational equation that was derived from the mathematical analysis of a kinetic model of the biochemical behaviour of [^{14}C]DG and glucose in brain (*Figure 1*). Although the model and its mathematical analysis were as rigorous and comprehensive as reasonably possible, it must be recognized that models always represent idealized situations and cannot possibly take into account every known, let alone unknown, property of a complex biological system.

The main potential sources of error are in the rate constants and the lumped constant. The problem is that they are not determined in the same animals and at the same time when local cerebral glucose utilization is being measured. They are determined in separate groups of comparable animals and then used, subsequently, in other animals in which glucose utilization is being measured.

8.1 Rate constants

The rate constants vary not only from structure to structure but with the condition. For example, K_1^* and k_2^* are influenced by both blood flow and transport of [^{14}C]DG across the blood–brain barrier, and because of the competition for the transport carrier, the glucose concentrations in the plasma and tissue affect the transport of [^{14}C]DG and, therefore, also K_1^* and k_2^*. The constant, k_3^*, is related to phosphorylation of [^{14}C]DG and can change when glucose utilization is altered. To minimize potential errors due to inaccuracies in the values of the rate constants used, a procedure was designed that sacrificed time resolution for the sake of accuracy. If the $[^{14}C]DG$ is administered as an intravenous pulse and sufficient time is allowed for the plasma to be cleared of the tracer by the body as a whole, then the influence of the rate constants, and the functions that they represent, on the final result diminishes with increasing time until eventually it reaches zero at infinite time (19). The use of $[^{14}C]DG$ instead of $[^{14}C]$ glucose permits the prolongation of the experimental period, perhaps, not to infinite times, but, at least, long enough for inaccuracies in the estimates of the rate constants to have relatively little effect on the final result. Because of this insensitivity to the values of the rate constants at such late times, for example 30–45 min, it is normally acceptable to use the average rate constants for grey and white matter given in *Table 1* rather than the individual rate constants for each region within the brain. Increases in the values of the rate constants, such as those that occur in hypoglycaemia because of the reduction in competitive inhibition by glucose, result in even lesser influence of the rate constants on the calculated values for glucose utilization (19). On the other hand, severe reductions in the values of the rate constants, such as those that might occur in pathological conditions, like severe ischaemia or hyperglycaemia, enhance their influence in the calculations (20). Values for the rate constants at various arterial plasma glucose levels ranging from normoglycaemia to severe hyperglycaemia are given in *Table 1*. In other conditions it may be necessary to redetermine the rate constants for the particular condition under study.

8.2 Lumped constant

The lumped constant is composed of six separate constants. One of these, ϕ, is a measure of the steady-state hydrolysis of glucose-6-phosphate to free glucose and phosphate. When there is no such hydrolysis, ϕ is equal to 1.0; it decreases toward zero with increasing degree of hydrolysis. Because there is little G-6-Pase activity in brain (1, 19, 21), ϕ is essentially equal to unity in the brain. In spite of considerable evidence that G-6-Pase activity is negligible in brain and is of no significance to the DG method if the experimental period is limited to 45 minutes (1, 19, 22), there have been a few reports that G-6-Pase activity is a major source of error (23–25). The basis of each of these reports has been carefully examined and shown to be flawed either as a result of misinterpretation of experimental findings or faulty biochemical procedures (26–28). The other components of the lumped constant are arranged in three ratios: λ, which is the ratio of distribution spaces in the tissue for deoxyglucose and glucose; and V_m^*/V_m and K_m/K_m^*, which are the ratios of the corresponding Michaelis–Menten kinetic constants of hexokinase for DG and glucose (*Figure 1*). Although each individual constant may vary from structure to structure and condition to condition, the ratios tend to remain relatively stable under normal conditions.

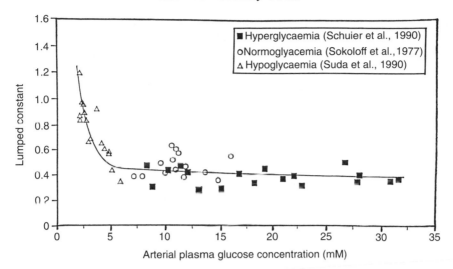

Figure 4. Values of lumped constant for various levels of arterial plasma glucose concentration from hypoglycaemia to hyperglycaemia. From Schuier *et al.* (30) based on data of Sokoloff *et al.* (1), Suda *et al.* (29), and Schuier *et al.* (30).

For reasons described in detail elsewhere (1, 19, 29, 30), the lumped constant tends to be the same throughout the brain and is characteristic of the species of animal, but only in normal tissue. Because of the slightly different kinetic behaviours of glucose and DG in the tissues, the lumped constant changes when the arterial plasma glucose content falls outside the normal physiological range, particularly in hypoglycaemia. In hypoglycaemia (i.e. arterial plasma glucose concentrations below 5 mM) values for the lumped constant rise sharply (29), and in hyperglycaemia the values tend to decline only slightly, even at arterial plasma glucose concentrations as high as 30–35 mM (*Figure 4*) (30). There is also reason to expect that the lumped constant could change in pathological conditions. Tissue damage may disrupt the normal cellular compartmentation, and λ, the ratio of the tissue distribution spaces for [^{14}C]DG and glucose, may not be the same in damaged tissue as in normal tissue.

8.3 Influence of varying plasma glucose concentration during the experimental period

Because the operational equation of the method (*Figure 1B*) was derived on the basis of the assumption that C_P, the arterial plasma glucose concentration, remains constant during the experimental period, the method was at first applicable only to conditions in which this assumption was valid. In most experimental conditions in which the DG method has been applied, plasma glucose concentration changes vary little in the course of the experiment. Excessive blood loss from sampling may lower blood pressure and raise plasma glucose content slightly by 20–30 minutes into the experimental period, but a 10–20% rise this late in the experimental period has little influence on the calculated rate of glucose utilization, even if the original operational equation is used. For conditions in which plasma glucose content changes more drastically from its initial value during the experimental period, an alternative equation has been developed, but it requires an estimate of the half-life of the free glucose pool in the brain tissue (31). The half-life for free glucose in brain was measured and found to be about 1.2 and 1.8 minutes in normal conscious and anaesthetized normoglycaemic rats, respectively, and to vary with the plasma glucose concentration (31). The equation is, however, relatively insensitive to the value of the half-life for the glucose pool in the range in which it usually falls, and, therefore, only an approximation of the half-life is sufficient without jeopardizing the accuracy of the final result. The [^{14}C]DG method can now, therefore, also be used in the presence of drastically changing arterial plasma glucose concentrations (31).

8.4 Animal behaviour during the experimental period

In order for the DG method to provide results representative of a given behavioural state, that state should be sustained over the entire 45 minutes of the experimental period. Behaviour lasting for only a short time, as in the performance of a brief task, is best represented when continuously repeated in a stereotyped manner. Seizures may be difficult to evaluate when the goal is to obtain accurate values for glucose utilization during the ictal episode. Focal motor seizures, manifested by repetitive jerking or twitching of the involved part of the body, such as those induced by intracortical injections of penicillin, may continue for long periods, but generalized convulsions often last only a few minutes. Even with recurrent seizures there may be intermittent periods of post-ictal depression. When there is such fluctuation in extremes of behavioural activity, the measured value for glucose utilization reflects a time-weighted average of the effects of all states of behaviour taking place during the experimental period. This relationship has been used in a study of rapid-eye-movement (REM) sleep (32). In this study the time occupied by the REM episodes during the experiment was only a fraction of the total 45 minute period and was interspersed between periods of non-REM sleep. The total time in each stage of sleep varied between experiments. The effect of REM versus non-REM sleep was evaluated by plotting the calculated rate of glucose utilization over the 45 minute period for each brain region against the per cent time spent in the given sleep stage, weighted for the estimated average free [^{14}C]DG content in the tissue that prevailed during that time interval of that sleep stage. It was thus possible to demonstrate that REM sleep increases the rate of glucose utilization in brain above that in non-REM sleep (32).

9. The use of [^{14}C]deoxyglucose in qualitative metabolic mapping of pathways and other non-quantitative studies

[^{14}C]DG is often used qualitatively to map functional pathways in the brain (33, 34). In this case the arterial plasma concentrations of [^{14}C]DG and glucose need not be monitored, and local glucose utilization is not calculated; the autoradiograms themselves provide a picture of relative local rates of glucose utilization within the brain. This is useful in studies in which a pathway is deprived or stimulated on one side of the brain, and the other side serves as the control (13, 33–37). Because local rates in laboratory animals are generally symmetrical under normal circumstances, findings of even small right-to-left differences in the autoradiograms may be of significance. It should be noted that the DG method results in labelling of entire functional pathways. This is illustrated in the demonstration of the ocular dominance columns in the monkey in studies of monocular occlusion; the reduced function in the neural pathway from the occluded eye is clearly evident beyond the laminae of the lateral geniculate bodies and beyond the terminals of the geniculostriate pathway in Layer IV of the striate cortex, extending to all cortical layers of the striate cortex (34).

Another non-quantitative use of [^{14}C]DG is in the localization of local cerebral activities in altered physiological states or drug-induced behaviours. Autoradiograms may show regional differences in optical density resulting in a pattern characteristic for the given state (15, 38–41). A difference in the pattern from that seen in normal resting animals is apparent by inspection. Quantitative interpretations of the images with respect to rates of glucose utilization by mere comparisons of optical densities or ^{14}C contents in any one region of the brain of one animal with that in the same region of another animal without employing the fully quantitative method is, however, unreliable. This is because the ^{14}C concentration in a given brain region is determined not only by the rate of glucose utilization in that region but also by the entire time courses of the arterial plasma [^{14}C]DG and glucose concentrations. It cannot be assumed that these time courses are the same in different animals simply because each is given equal doses of [^{14}C]DG per kg of body-weight. Not only are there normal random physiological variations from animal to animal that affect these time courses, but there may be systematic alterations characteristic of the given experimental condition. In non-quantitative as well as quantitative versions of the DG method it is best to administer the [^{14}C]DG intravenously as a pulse if possible. Subcutaneous and intra-peritoneal routes of administration have been used, but with some loss of metabolic discrimination and anatomical detail in the autoradiograms because of the higher fraction of the total ^{14}C in the form of unmetabolized [^{14}C]DG. In studies in insect brain, the oral route was found to be satisfactory (36).

A number of reports have appeared in which radiolabelled DG was employed in a 'semi-quantitative' procedure. This consists of administering the DG, removing and freezing the brain at the prescribed time, and sectioning and processing the tissue for autoradiography. Optical densities or local isotope concentrations are measured in the regions of interest, they are then reported as ratios normalized to the corresponding average values determined in some arbitrarily chosen white matter structures where it is assumed that glucose utilization is unaffected by the experimental condition. These ratios are presented as indices of the rates of glucose utilization. This approach may occasionally be useful in comparing metabolic rates, provided that it is first demonstrated that glucose utilization is indeed unaffected in the structures used as the references under the experimental conditions (42). In most reports, however, this has not been established, and the results are, therefore, of uncertain significance. The rationale often presented to justify this abridged procedure, resulting in an index rather than absolute values for glucose utilization, is the need to have the animal unrestrained and 'unstressed'. Inasmuch as blood-sampling catheters can readily be arranged to permit behavioural studies to be conducted in freely-moving animals that show no evidence of stress (8, 43, 44), there seems little justification to abandon the fully quantitative method and obtain actual rates of glucose utilization.

10. Microscopic resolution

The spatial resolution of the $[^{14}C]DG$ method is approximately 100–200 µm (45). The use of $[^{3}H]DG$ does not significantly improve the resolution with autoradiography of 20 µm thick brain sections because the limiting factor is the diffusion of the water-soluble labelled compounds during the freezing and cutting of the brain sections (45). In order to significantly improve the level of resolution it is necessary to employ procedures for fixation of the label *in situ* so that it cannot migrate during the freezing and cutting of the brain nor be washed out at any time during the processing of the tissue and the autoradiography. Unfortunately, both conventional fixation with glutaraldehyde (46) or with periodate, lysine, and paraformaldehyde (47) result in large (90%) losses of label from the tissues. Chemical fixation of the label *in vivo* is the desired approach to this problem, but a practical procedure for fixation of sugars *in vivo* has not yet been achieved. Alternative solutions are to remove all water from the tissue either by freeze-drying (36, 48) or freeze-substitution (49, 50) or to carry out all procedures, including autoradiography, on frozen tissues (51). These techniques have been used successfully at the electron microscopic level to map odour-induced neuronal activity in the olfactory bulb of the rat (52) and the effects of visual stimulation on neuronal activity in drosophila (53).

11. Double-label autoradiography

DG labelled with ^{14}C and with either ^{3}H or ^{18}F on the 2-carbon position has been used in sequential double-label procedures to compare two states of behaviour or functional activation in the same animal (54–58). Although some success has been achieved in the separation of the metabolic effects associated with the two different experimental conditions, technical problems remain to be solved before valid quantitative measurements of local cerebral glucose utilization can be obtained in such sequentially conducted studies. One problem is the differential self-absorption of the weak β-radiation of ^3H in different structures of the brain. The self-absorption is greater in white matter than in grey matter (59), presumably because of the greater lipid content in white matter. Therefore, because the proportions of grey and white matter differ among the various structures of the brain, the optical densities in autoradiograms made with ^3H do not accurately reflect local ^3H concentrations. A second problem is that significant amounts of unmetabolized DG labelled with the first isotope still remain in the blood and tissues at the end of the first experimental period, and it is then metabolized in the second experimental period according to the metabolic rates that prevail during those experimental conditions. The amounts are sufficient to distort the results of the first period. Some of these problems have recently been reviewed (57), but until they are satisfactorily resolved the quantitative DG method is best used to compare experimental conditions in single procedures in separate groups of animals.

References

1. Sokoloff, L., Reivich, M., Kennedy, D., Des Rosiers, M. H., Patlak, C. S., Pettigrew, K. D., Sakurada, O., and Shinohara, M. (1977). *J. Neurochem.*, **28**, 897.
2. Nelson, T., Kaufman, E. E., and Sokoloff, L. (1984). *J. Neurochem.*, **43**, 949.
3. Kennedy, C., Sakurada, O., Shinohara, M., Jehle, J., and Sokoloff, L. (1978). *Ann. Neurol.*, **4**, 293.
4. Duffy, T. E., Cavazzuti, M., Cruz, N. F., and Sokoloff, L. (1982). *Ann. Neurol.*, **11**, 233.
5. Abrams, R. M., Ito, M., Frisinger, J. E., Patlak, C. S., Pettigrew, K. D., and Kennedy, C. (1984). *Am. J. Physiol.*, **246**, R608.
6. Reivich, M., Alavi, A., Wolf, A., Fowler, J., Russell, J., Arnett, C., MacGregor, R. R., Shiue, C. Y., Atkins, H., Anand, A., Dann, R., and Greenberg, J. H. (1985). *J. Cereb. Blood Flow Metab.*, **5**, 179.
7. Nehlig, A., de Vasconcelos, A. P., and Boyet, S. (1988). *J. Neurosci.*, **8**, 2321.
8. Crane, A. M. and Porrino, L. J. (1989). *Brain Res.*, **499**, 87.
9. Patlak, C. S. and Pettigrew, K. D. (1976). *J. Appl. Physiol.*, **40**, 458.
10. Wree, A., Schleicher, A., Zilles, K., and Beck, T. (1988). *Histochemistry*, **88**, 415.
11. Goochee, C., Rasband, W., and Sokoloff, L. (1980). *Ann. Neurol.*, **7**, 359.
12. Sokoloff, L. (1981). *J. Cereb. Blood Flow Metab.*, **29**, 13.
13. Macko, K. A., Jarvis, C. D., Kennedy, C., Miyaoka, M., Shinohara, M., Sokoloff, L., and Mishkin, M. (1982). *Science*, **218**, 394.
14. Newman, G. C., Hospod, F. E., and Patlak, C. S. (1990). *J. Cereb. Blood Flow Metab.*, **10**, 510.
15. Schwartz, W. J., Smith, C. B., Davidsen, L., Savaki, H., Sokoloff, L., Mata, M., Fink, D. J., and Gainer, H. (1979). *Science*, **205**, 723.

16. Mata, M., Fink, D. J., Gainer, H., Smith, C. B., Davidsen, L., Savaki, H., Schwartz, W. J., and Sokoloff, L. (1980). *J. Neurochem.*, **34**, 213.
17. Kadekaro, R. M., Crane, A. M., and Sokoloff, L. (1985). *Proc. Natl. Acad. Sci. USA.*, **82**, 6010.
18. Soncrant, T. T., Holloway, H. W., Stipetic, M., and Rapoport, S. (1988). *J. Cereb. Blood Flow Metab.*, **8**, 720.
19. Sokoloff, L. (1982). In *Advances in neurochemistry* (ed. B. W. Agranoff and M. H. Aprison), Vol. 4, pp. 1–82. Plenum Press, New York.
20. Orzi, F., Lucignani, G., Dow-Edwards, D., Namba, H., Nehlig, A., Patlak, C. S., Pettigrew, K., Schuier, F., and Sokoloff, L. (1988). *J. Cereb. Blood Flow Metab.*, **8**, 346.
21. Hers, H. G. (1957). *Le métabolisme du fructose*, p. 102. Editions Arscia, Bruxelles.
22. Mori, K., Schmidt, K., Jay, T., Palombo, E., Nelson, T., Lucignani, G., Pettigrew, K., Kennedy, C., and Sokoloff, L. (1990). *J. Neurochem.*, **54**, 307.
23. Hawkins, R. A. and Miller, A. L. (1978). *Neuroscience*, **3**, 251.
24. Huang, M.-T. and Veech, R. L. (1982). *J. Biol. Chem.*, **257**, 11358.
25. Sacks, W., Sacks, S., and Fleischer, A. (1983). *Neurochem. Res.*, **8**, 661.
26. Nelson, T., Lucignani, G., Atlas, S., Crane, A. M., Dienel, G. A., and Sokoloff, L. (1985). *Science*, **229**, 60.
27. Nelson, T., Lucignani, G., Goochee, J., Crane, A. M., and Sokoloff, L. (1986). *J. Neurochem.*, **46**, 905.
28. Dienel, G. A., Nelson, T., Cruz, N. F., Jay, T., Crane, A. M., and Sokoloff, L. (1986). *J. Biol. Chem.*, **263**, 19697.
29. Suda, S., Shinohara, M., Miyaoka, M., Lucignani, G., Kennedy, C., and Sokoloff, L. (1990). *J. Cereb. Blood Flow Metab.*, **10**, 499.
30. Schuier, F., Orzi, F., Suda, S., Lucignani, G., Kennedy, C., and Sokoloff, L. (1990). *J. Cereb. Blood Flow Metab.*, **10**, 765.
31. Savaki, H. E., Davidsen, L., Smith, C., and Sokoloff, L. (1980). *J. Neurochem.*, **35**, 495.
32. Abrams, R. M., Hutchison, A. A., Jay, T. M., Sokoloff, L., and Kennedy, C. (1988). *Dev. Brain Res.*, **40**, 65.
33. Kennedy, C., Des Rosiers, M. H., Jehle, J. W., Reivich, M., Sharp, F., and Sokoloff, L. (1975). *Science*, **187**, 850.
34. Kennedy, C., Des Rosiers, M. H., Sakurada, O., Shinohara, M., Reivich, M., Jehle, J. W., and Sokoloff, L. (1976). *Proc. Natl. Acad. Sci. USA*, **73**, 4230.
35. Collins, R. C., Kennedy, C., Sokoloff, L., and Plum, F. (1976). *Arch. Neurol.*, **33**, 536.
36. Buchner, E., Buchner, S., and Hengstenberg, R. (1979). *Science*, **205**, 687.
37. McCulloch, J., Savaki, H. E., McCulloch, M. C., and Sokoloff, L. (1980). *Science*, **207**, 313.
38. Pulsinelli, W. A. and Duffy, T. E. (1978). *Science*, **204**, 626.
39. Hubel, D. H., Wiesel, T. N., and Stryker, M. P. (1978). *J. Comp. Neurol.*, **177**, 361.
40. Kliot, M. and Poletti, C. E. (1979). *Science*, **204**, 641.
41. Meibach, R. C., Glick, S. D., Cox, R., and Maayani, S. (1979). *Nature*, **282**, 625.
42. Toga, A. W. and Collins, R. C. (1981). *J. Comp. Neurol.*, **199**, 443.
43. Porrino, L. J., Esposito, R. U., Seeger, T. F., Crane, A. M., Pert, A., and Sokoloff, L. (1984). *Science*, **224**, 306.
44. Jay, T. M., Jouvet, M., and Des Rosiers, M. H. (1985). *Brain Res.*, **342**, 297.

45. Smith, C. B. (1983). In *Current methods in cellular neurobiology*, Vol. I. *Anatomical techniques*. (ed. J. L. Barker and J. F. McKelvy), pp. 269–317. John Wiley, New York.
46 Ornberg, R. L., Neale, E. A., Smith, C. B., Yarowsky, P., and Bowers, L. M. (1979). *J. Cell. Biol. Abstr.*, **83**, 142A.
47. Durham, D., Woolsey, T. A., and Krugher, L. (1981). *J. Neurosci.*, **1**, 519.
48. Duncan, G. E., Stumpf, W. E., and Pilgrim, C. (1987). *Brain Res.*, **401**, 43.
49. Sejnowski, T. J., Reingold, S. C., Kelley, D. B., and Gelperin, A. (1980). *Nature*, **287**, 449.
50. Lancet, D., Greer, C. A., Kauer, J. S., and Shepherd, G. M. (1982). *Proc. Natl. Acad. Sci. USA*, **79**, 670.
51. Hökfelt, T., Smith, C. B., Peters, A., Norell, G., Crane, A., Brownstein, M., and Sokoloff, L. (1983). *Brain Res.*, **289**, 311.
52. Benson, T. E., Burd, G. D., Greer, C. A., Landis, D. M. D., and Shepherd, G. M. (1985). *Brain Res.*, **339**, 67.
53. Buchner, S. and Buchner, E. (1982). *Neurosci. Lett.*, **28**, 235.
54. Altenau, L. L. and Agranoff, B. W. (1978). *Brain Res.*, **153**, 375.
55. Olds, J. L., Frey, K. A., Ehrenkaufer, R. L., and Agranoff, B. W. (1985). *Brain Res.*, **361**, 217.
56 Juhler, M. and Diemer, N. H. (1987). *J. Cereb. Blood Flow Metab.*, **7**, 572.
57. Redies, C. and Gjedde, A. (1989). *Cerebrovasc. Brain Metab. Rev.*, **1**, 319.
58. Friedman, H. R., Bruce, C. J., and Goldman-Rakic, P. S. (1987). *Exp. Brain Res.*, **66**, 543.
59. Orzi, F., Kennedy, C., Jehle, J., and Sokoloff, L. (1983). *J. Cereb. Blood Flow Metab.*, **3**(Suppl. 1), S77.
60. Sokoloff, L. (1985). *The Harvey lectures*, **79**, 77.

9

Fluorescence—
monitoring cell chemistry *in vivo*

MICHAEL R. DUCHEN

1. Introduction

Many molecules, when exposed to light of appropriate energy, absorb photons which excite electrons into a higher energy state. As those electrons relax to their original orbits, photons are re-emitted, a phenomenon known as fluorescence. The detailed spectra of the light that is absorbed and emitted may vary accordinging to other interactions of the molecule—its binding to Ca^{2+}, protons, or other ions, the electrical field it senses, and so on. In recent years, we have seen major developments in fluorescent probes that can transmit information about intracellular events within living cells. The changes in intracellular concentrations of ions and other second messenger systems that play so fundamental a role in cell function have at last become accessible to direct study without having to impale cells with microelectrodes.

Superficially, light seems an innocuous, non-invasive tool that makes possible the study of a whole range of variables in living cells with a minimum of interference. One might expect that many probes, and certainly intrinsically fluorescent compounds within the cell (see below), would simply report changes in the variables under study without any significant influence on cell function. The reality may be less straightforward as some probes turn out to be toxic or biologically active, and even the energy of the light used to excite the fluorescence may itself be sufficient to alter cell behaviour. These problems must all be recognized and appropriate care taken to avoid them. Nevertheless, the technology is developing so rapidly that this is a most exciting time to be engaged in such work—almost anything one chooses to study is likely to prove new and instructive.

1.1 Fluorescence vs absorbance and chemiluminescence

Fluorescent probes have largely superseded their forerunners, such as the Ca^{2+} sensitive probes, aequorin and Arsenazo, which were not fluorescent. Aequorin is a chemiluminescent photoprotein that emits photons on binding Ca^{2+}, while Arsenazo III has an absorption spectrum which alters on binding Ca^{2+}.

Fluorescent indicators have significant advantages over both of these. A chemiluminescent probe allows only the release of a single photon for each molecular interaction that takes place, after which the molecule of dye is 'used up', while a fluorescence probe can undergo millions of transitions. Fluorescence signals are typically detected against a background which has negligible emission, so that the signal to noise ratio can be kept high. In contrast, detection of a change in absorption may require detection of a relatively small proportional change in light intensity (especially so in small cells or processes) against a relatively bright background. There are other problems associated with Arsenazo III that are not relevant to this discussion, but few would now choose to use either system given the choice of the new fluorescent probes.

2. What can be measured?

The earliest attempts to measure light to illuminate biological problems involved study of phenomena such as the rather subtle changes in the birefringency of neurons during their electrical activity (1); the generation of chemiluminescence of naturally occurring proteins in response to changes in $[Ca^{2+}]$ or ATP (2); and changes in the absorption spectra of biochemically active pigments (3). Thanks to numerous reviews and papers in prominent journals, most investigators are now aware that it is possible to monitor changes in intracellular calcium concentration, $[Ca^{2+}]_i$, continuously with a number of different fluorescent probes. A change in $[Ca^{2+}]_i$ is such an important and ubiquitous signal that this is given much attention below. But, paging through the catalogues from Molecular Probes (Oregon, USA) or Calbiochem (Nottingham, UK) for example, it will soon become obvious that there are large numbers of probes available to measure other variables which have barely been exploited. It is currently feasible to use fluorescence technology to monitor $[Ca^{2+}]_i$, pH_i, $[Na^+]_i$, $[Mg^{2+}]_i$, cell membrane potential, $[Cl^-]_i$, $[Zn^{2+}]_i$, [NADH/NAD] ratio, mitochondrial membrane potential, and to monitor secretion (see also ref. 4). Probes are available to follow the progress of biochemical reactions inside the cell. The ionic probes are not all ideal—some are not as specific as might be desired—but it seems inevitable that further probes will become available in the next few years, driven by the development of a technology which is so powerful once applied.

The introduction of the patch clamp technique and its combination with microfluorimetry, makes possible investigation of the interrelationships between variables such as $[Ca^{2+}]_i$, pH_i etc., the modulation of membrane conductances, the consequences of the introduction of biologically interesting compounds into the cytoplasm, and so on. This represents an opportunity which has barely been tapped in the study of the CNS.

2.1 Levels of resolution

2.1.1 Spectrofluorimetry

A spectrofluorimeter may be used to look at populations of cells or subcellular particles. The techniques have been mostly applied to cells which are easily obtained in bulk and which readily withstand isolation to remain viable *in vivo*—neutrophils, platelets, hepatocytes, mast cells, etc.—or on synaptosomes (5), and isolated organelles. The instrument monitors fluorescence from material in a cuvette, usually from cells or fragments which are suspended in the cuvette. Occasionally cells studied this way are grown on a coverslip which is then placed in the cuvette or even grown in culture on the side of the cuvette. Once a drug is added to the suspension it cannot usually be removed, so that most experiments consist of a few sequential manipulations after which the suspension is discarded. This is fine if there is an abundant supply of cells or synaptosomes, but in studies of whole mammalian neurons, this is rarely the case.

2.1.2 Fluorescence microscopy

This technique is used to measure signals from single cells or small clusters of cells. The fluorescence signal from a given microscopic field is measured, representing the average emission from all components of the field. Again, application of this (and 2.1.3 and 2.1.4) to the study of mammalian neurons has been limited by their inaccessibility. Brain slice preparations are the most complex used for these studies, and still have many problems, although recent developments (6, 7) will make their use more widespread. Others, including myself, have used preparations of neurons freshly dissociated from the CNS and other tissues (8), while cells (usually neonatal) grown in culture have perhaps been more widely used. Cells are usually adherent to a glass or quartz coverslip, and can be continuously superfused so that the effects of numerous manipulations may be studied sequentially on a single cell. As the cells are directly visualized, healthy cells (readily identified under phase contrast or Nomarski optics) or cells of a certain shape can be selected. The technique allows simultaneous electrophysiological study. These advantages apply equally to (2.1.3) and (2.1.4) below. As this is the area of my own experience, it is in this area that I will focus below. For detailed discussions of the principles of fluorescence microscopy, see reference 9.

2.1.3 Fluorescence imaging

Fluorescence imaging is an elaboration of fluorescence microscopy that enables the two dimensional *distribution* of fluorescence to be studied. It allows discrimination of changes in signal in different parts of the field of view. This could include simultaneous study of several cells in a field, or of various components and processes of a single complex cell. The computing required to do this is complex, and the potential artefacts and errors in interpretation of the data are many, but the techniques are powerful and have already provided valuable insights.

2.1.4 Con-focal scanning laser microscopy

This is a further development that improves spatial resolution and allows the study of fluorescence of elements within a complex tissue. This is the most expensive application, only recently developed, and promises to be enormously valuable.

3. Basic requirements

3.1 Elimination of vibration

For applications 2.1.2–2.1.4, you will need to reduce the transmission of vibration to the system, especially if the measurement of fluorescence is coupled with electrophysiological recording. The requirements vary considerably between buildings (see also Chapter 3). We have used a metal or slate baseplate supported on either tennis balls or inner tubes (for a small car wheel) on a heavy table. The best solution is unquestionably a purpose made anti-vibration table, which supports a table base on gas-filled pistons. A range of tables can be bought from suppliers of optical equipment (Newport Corp, Oriel Corp., Wentworth Laboratories Ltd. We use a micro-G table from Technical Manufacturing Corp. MA, USA).

3.2 The light path

Most microfluorimetric systems are based around an inverted microscope, in which the working space between the condenser and the preparation greatly facilitates all manipulations. The microscope should be equipped with phase contrast, Hoffman, or Nomarski differential interference optics to aid visualization of cells. Our system is based on the Nikon Diaphot microscope, as are several of the commercially available packages detailed at the end of this chapter. It is important that the microscope does not use too many glass components which exclude the use of UV light and which often show significant autofluorescence. *Figure 1* illustrates the essential features of our own system as arranged for using the $[Ca^{2+}]_i$ sensitive probe, Indo-1. The system includes elements built in our mechanical workshop and added to the original microscope. Several components are now available from Nikon, including beam splitters and diaphragms.

The intensity of the fluorescence signal from a probe carries information about the variable to which it is sensitive but it is also a function of the concentration of the probe. Thus, signals cannot be directly calibrated simply from the measurement of fluorescence intensity. The problem has been recently overcome with the introduction of dual emission and dual excitation probes, for which the response to the variable studied is a shift in emission or excitation wavelength,

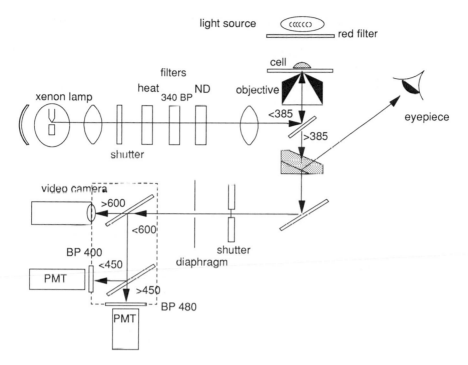

Figure 1. The arrangement of our optics for measurement of the dual emission dye, Indo-1. The correct alignment of the beam-splitting dichroic mirrors is critical to optimize the projection of the image on to the face of the PMT. The 340 nm bandpass (BP) filter is actually fitted into a block that can rotate to present either of two filters in the light path. The set-up can thus be used for either dual excitation or dual emission measurements. The dotted line shows the enclosed box that we have made ourselves.

rather than a shift in intensity. Calibration is then a function only of the relative position of the spectra, independent of dye concentration. This will be discussed at greater length below (Section 7), but it does explain why most would advocate the design of a system which enables both switching between two excitation wavelengths and measurement of fluorescence emission at two wavelengths, such as the system illustrated.

3.3 Choice of light source

Mercury or xenon arc lamps are used for most applications. Short arc lamps are the brightest available sources of light with the exception of lasers. The light is generated by a high voltage arc through a gas. The arc between anode and cathode is so short that it acts as a point source. The anode and cathode are made of tungsten doped with various elements to improve electron emission, enclosed in a quartz envelope filled with either a rare gas at several atmospheres of pressure or with a little rare gas and an exact amount of mercury.

- *Mercury lamps* contain an exactly measured amount of Hg which is vaporized by the high temperatures. The excited Hg atoms emit bands of light from the UV to the visible range. Individual lines are selected using mono-chromators or filters.

- *Xenon lamps* are filled with purified xenon at 5–20 bar. Through the UV and visible ranges, the spectrum of light emitted is reasonably flat, so that these sources are more flexible. Xenon lamps last longer than mercury, saving in the cost of replacements. 75–150 W bulbs are used in most micro-fluorimetric applications.

- *Quartz tungsten halogen lamps* provide long wave visible to infrared illumination. The requirements for such wavelengths are limited in the applications discussed here.

- Occasionally, a *laser light* source is used. This has the great advantage of high intensity and uniformity of wavelength, but is expensive and less flexible. A prime application of laser light now is in con-focal scanning micro-fluorimetry, or in flash photolysis of caged compounds, where the high energy and focus of the beam is a major advantage.

3.3.1 Fluctuations in light intensity

Fluctuations in light intensity may prove indistinguishable from fluctuations in the biological variable. Fluctuations in ambient light are probably the major problem, and can be avoided by making the system light tight or by working in the dark. The light source must have a stabilized DC power supply, which will supply a constant current or power as the line voltage varies. It is possible to buy feedback regulators which have a photodiode immediately in front of a light source, and a circuit to adjust the power supply to correct any detected fluctuation, but most people manage without these.

Some spectrofluorimeters are designed to overcome fluctuations in light intensity by splitting the light beam into two, one of which bypasses the pre-paration. This is continuously monitored as a reference signal, with which the signal from the preparation is compared. Fluctuations from the light source can then be subtracted from the signal from the preparation. This could be done in dual emission microfluorimetry (see below) for single wavelength dyes. Monitor both the main emission peak and a longer wavelength where the emission is insigni-ficant, and where variations in light should be due only to fluctuations in the source.

One source of fluctuation may be within the light itself, when the light switches as though between two states. It seems that the arc alternates between two routes between anode and cathode. Occasionally, this may be a transient phenomenon, and the source settles down, presumably to the favoured arc pathway, but usually it means changing the bulb.

3.3.2 Problems

UV and infra-red light damage cells. A beam of UV light focused intensely on to a single cell may kill it or impair its function—so-called photodynamic damage.

Some of the damage is due to the production of oxygen free radicals by the UV light, and may be reduced by adding antioxidant to the salines. This is not appealing, however, and it is better to limit the intensity of light to the minimum that will give a reasonable signal (that also means optimizing the collection of the signal and its detection). This is best achieved by placing heat filters, diffusing filters, and neutral density filters in the light path. The final selection used is empirically determined by choosing filters that give a reasonable trade-off between loss of signal (and therefore, decreasing signal/noise ratio) on the one hand and deterioration of the preparation on the other. We have a range of neutral density filters with transmittance from 33% down to 1%.

Some spectrofluorimeters use intermittent pulses of light to reduce the exposure of cells to UV light, and a similar approach can be used in microscopy to study slow processes (for example over tens of minutes). Light can be simply excluded with an electrically operated shutter (for example UniBlitz, Vincent Associates, Rochester, USA) that can be triggered to open or close by a TTL pulse generated from a computer, Digitimer modules or other source.

3.3.3 SAFETY

Arc lamps are under high pressure even when cold. They should be handled with great respect, using goggles and gloves at the very least. The quartz envelope must not be touched as contamination increases thermal stress. The light source must only be operated in an enclosed housing. During operation, the temperature inside the lamp rises to about 2000 K, and the pressure to 20–75 bar. They are POTENTIALLY DANGEROUS, occasionally exploding with a force sufficient to shatter a glass lens.

UV radiation damages eyes. In the short term, very painful irritation of the conjunctiva results, which may not be apparent until several hours after exposure to light at wavelengths shorter than 400 nm, while long-term exposure to UV light can cause lenticular cataract. **Always** use protective goggles when examining the light paths, etc, to limit exposure, and make sure other people in the vicinity are similarly protected. Place filters as close as possible to the light source if UV light is not needed for the experiment (but also remember that the source becomes hot, and that the filters may change their characteristics when hot).

Ozone is generated by lamps which produce light at wavelengths below 200 nm. It is toxic, causing irritation to mucous membranes and to the eyes. One solution: use ozone-free lamps which have an envelope that absorbs light below 300 nm. Very few applications in biology require light at such short wavelengths. Extraction systems are commercially available to remove ozone from the lamp housing.

3.4 Selection of wavelengths

Either monochromators or optical filters are used to select the wavelength needed to provoke excitation of the fluorescence probe or to be measured by the detection system.

3.4.1 Monochromators

The central element is a diffraction grating that is moved relative to the light source to vary the wavelength selected to fall on an output slit. These are used by most spectrofluorimeters. For microfluorimetry, they represent an expensive option. They do have the advantage of flexibility, as different wavelengths may be chosen easily, rather than having to buy a new optical filter each time you want to change the excitation wavelength. They occupy significant space on a set-up, the focusing of the input beam must be optimized, and most have a small exit slit which makes the focusing of the excitation beam more critical. All but the more expensive lose significant light energy at high resolution. The monochromator gives the ability to measure full spectra as the wavelength can be continuously varied, and can be driven by computer, which would obviously be an attraction for some experiments. Some investigators have opted to use monochromators on both the excitation and emission sides of a micro-fluorimetric system to give maximum flexibility, but such systems are clearly more important to those developing new applications than to most users. Such a system is described in some detail in ref. 10.

3.4.2 Optical filters

For microfluorimetry most people choose to use optical filters. These have the merit of a large surface area, small size, and low cost. They occupy little space on the set-up, and are very easy to install. In practice, we have found that for microscopy, by investing in a range of bandpass (± 5 or 10 nm) filters and combinations of short and long pass filters between 340–650 nm, we now usually have filters to meet most requirements for selection of either excitation or emitted light.

Special precautions
Optical filters are damaged by heat—both the filtering properties and the cement that holds it together. Most filters, therefore, have one reflective surface, which should always be positioned facing the light source. Excitation filters placed close to an arc lamp will get hot. Ideally, the filter should be positioned before any focusing mechanism, so that the beam is relatively diffuse. This will lessen the heat-induced deterioration of the filter. While heating reversibly shifts the selectivity to longer wavelengths, this amounts to only about 0.003% per °C, usually insignificant for the purposes of biological experiments. Prolonged heating also produces irreversible shifts in selectivity to shorter wavelengths, but again this tends to be small.

3.4.3 Further considerations on the selection of wavelengths used

Selection of the wavelength to be measured should be carefully considered in relation to the spectral characteristics of the probe. Narrow width bandpass filters (usually ± 5 nm) are needed to select the wavelengths used for a dual emission dye such as Indo-1, for example (see below). If the wavelength shift between excitation and emission is short, it is obviously important to exclude the exciting light and to select only the emitted light. However, if only one wavelength is monitored, and the excitation and emission spectra are reasonably far removed, then a wider bandpass filter or even a combination of short and long pass filters might be appropriate. Maximize the light collection, and you can then afford to reduce the exposure of the preparation to harmful light or increase the sensitivity of the system to look at subtle changes in fluorescence.

Figure 1 shows how we use a series of low pass dichroic mirrors to split the light beam emitted from the cells, first to a red-sensitive TV camera, and then to two photomultiplier tubes (PMT). The requirement for such a system should become clear below.

3.5 Beam splitting—separation of exciting and emitted light

The exciting beam of light is focused on to the preparation after reflection off a dichroic mirror (*Figure 1*). This is a long pass filter chosen to reflect the short wavelength light used for excitation of the preparation, and to allow through the longer wavelength light emitted as fluorescence by the preparation. In other words, the cut-off wavelength should lie between the excitation and emission wavelengths used. Thus, for Indo-1, with an optimal excitation wavelength of 340–350 nm, and two optimal emission wavelengths of 405 nm and 480 nm (see below), a dichroic mirror that reflects below 385 nm and transmits above 385 nm should be ideal. For Fluo-3, which is excited at about 500 nm and emits maximally at about 530 nm, a dichroic mirror that reflects below 510 nm and transmits above 510 nm is appropriate. For many applications, standard dichroic blocks are available. It may prove necessary to buy a dichroic mirror and an empty filter block. The spectrum of a typical long pass filter (dichroic) is shown in *Figure 2*. Note that reflection of light is effective only for a band of wavelengths, while at shorter wavelengths there is some transmission. The cut-off operates over about 10 nm. Thus, the filter should have a cut-off ideally about 20 nm longer than the excitation wavelength selected—if it is either too close or too distant, some light will probably be transmitted. Some manufacturers will make dichroics 'extended left', increasing the efficiency of reflection at shorter wavelengths.

4. Collection and detection of emitted light

The light emitted by the preparation is collected by a series of lenses and focused on to a light sensitive device, usually a PMT or, in the case of imaging applications, a silicon intensified target (SIT) or charge coupled device (CCD) video camera (see below).

4.1 Lenses

The most efficient collection of emitted light is achieved using a high quality objective lens with the highest available numerical aperture. Oil or glycerine

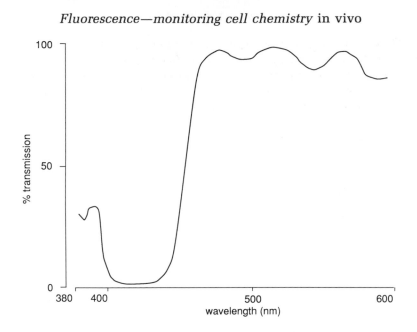

Figure 2. The spectrum supplied with the 450 nm dichroic mirror that we use as a beam splitter in our system, typical of the transmission spectra for most dichroics.

immersion lenses also help to minimize loss of signal. Glass absorbs light at wavelengths shorter than about 350 nm, so that if UV light is used, a quartz lens increases the efficiency of transmission. The most suitable of such lenses at present seem to be the Zeiss Neofluar ×63, 1.25 numerical aperture or the Nikon UV-F ×40, 1.3 numerical aperture. Even at longer wavelengths, signals collected through these lenses will have very much greater resolution than a glass lens of similar magnifying power.

4.2 Restriction of field

It is important in microfluorimetry to restrict the microscopic field projected to the detection system, as fluorescence is effectively averaged over the whole field arriving on the face of a PMT, and a significant contribution from the diffuse background autofluorescence of the optics and from cells which are not under study will decrease the resolution of the system. Restriction of the field is easily achieved by interposing a pinhole or variable diaphragm on the output arm of a fluorescence microscope.

4.3 Photomultiplier tubes (PMTs)

For non-imaging spectrofluorimetry, PMTs are about 100 times more sensitive to UV and visible light than are other detectors and they have suitably rapid responses.

PMTs vary in their spectral sensitivity depending primarily on the glass used for the window. Most manufacturers illustrate the spectral sensitivities of their PMTs, so that it is reasonably easy to choose one that is suitable for your application. For Indo-1 or Fura-2, the $[Ca^{2+}]_i$-sensitive probes, for example, measurement at 400–500 nm is achieved by most standard PMTs, while the pH sensitive dye, SNARF-1 and a number of voltage-sensitive dyes require measurements at long wavelengths, and may require special PMTs sensitive into the red range.

4.3.1 Light measurement with a PMT

A PMT operates with a high voltage supply that can be varied. When a PMT is supplied, information is provided about its spectral sensitivity and the optimal voltage range over which it should be operated. A dual emission system operates most efficiently if each PMT is driven at its own optimal voltage. We use a single power supply whose output is split between the two PMTs. When ordering PMTs with such a system, it is important to specify that they should be well matched, so that a voltage suitable for each can be used. The alternative is to use separate power supplies.

A PMT is damaged by exposure to bright light. It should, therefore, be protected from surges of light (for example, when changing filters, turning on the room lights, etc). This can be simply done by switching off the PMTs, but is also conveniently done by placing an electronic shutter in front of the PMT which is normally open, but can be closed with a TTL pulse, for example, whenever there is a risk of a light surge. Alternatively, the input can simply be blanked off manually with a sliding barrier.

4.3.2 Current/voltage converters vs photon counting

There are two ways of measuring light with a PMT. The current generated by the light falling on the photocathode can be directly monitored by using a current to voltage converter. Alternatively, the PMT may be used in photon counting mode. This method uses a pulse height discriminator which selects pulses of current from the background and converts these to TTL pulses. This signal can be input directly to a counter to count the number of pulses per unit time, effectively integrating the response over that time. My own experience is with a homemade current to voltage converter, and I have experienced no problems with the method.

However, photon counting is advocated by many on the grounds that (i) the signal/noise ratio may be improved, (ii) the output is digital, making subsequent on-line computing easy and avoiding the need for analog to digital conversion. Some of the commercially available microscofluorimetric systems offer an option on these methods.

Each method has its proponents who insist that only one method is worth using. Such advocacy usually means that the difference is not great, or there would be no discussion. Most fluorescence signals are reasonably bright, orders of magnitude greater than the dark noise of the PMT, so that in practice, there is little difference in the signals obtained. The main choice may lie in the data processing: photon counting is probably the choice if direct on-line computer storage is required. Photon counters are more expensive, as a simple current-to-voltage converter can quite easily be built in the lab. The background, or dark, noise of a PMT can also be reduced by using a cooled housing. For most applications, these are not required, and they increase the cost, but they do help to increase the sensitivity of the light detection.

To improve the signal/noise ratio and to protect the PMT it is necessary to reduce background or stray light to an absolute minimum. Rather than working in the dark, most choose to enclose the microscope and all attachments within a light tight housing, including covering the eyepieces. This is a problem if the experiments include manipulations such as patch clamping, in which case a compromise between a darkened room and limited light exposure of the microscope must be sought, perhaps with red darkroom style filters on lights used.

5. Artefacts

In some ways, consideration of the various sources of artefactual responses helps to explain the strategies that have been adopted specifically to avoid them. Consider the measurement of a signal from a probe where the spectrum remains constant with the variable measured, simply varying in the intensity of fluorescence emitted.

(a) As mentioned in Section 3.2 the primary problem with such measurements lies in calibration. It is impossible to differentiate the intensity of fluorescence due to changes in dye concentration or due to changes in the variable studied.

(b) A drug applied to a cell may be fluorescent itself at the wavelengths used, which makes it almost impossible to interpret the result of the experiment. As many compounds show some fluorescence, this is a significant source of artefact. Some drugs may interact with the dye in other ways, changing the light absorption, quenching fluorescence, or altering fluorescence through energy transfer (for example, a drug absorbs light of the same energy as that emitted by the probe, reducing the fluorescence signal from the probe).

(c) Many fluorescent probes show bleaching—a progressive decline in fluorescence intensity on exposure to light—making it impossible to distinguish a decrease in fluorescence of biological relevance or due to bleaching unless it is readily reversible. The same is true of changes in signal due to changes in focus of the preparation or movement of a cell, for

242

example in response to pressure ejection of a drug. We have found that changing the temperature may alter the configuration of our bath, altering the focusing of the emitted light on to the PMT, and therefore changing the apparent signal intensity. Again, if this is not borne in mind, the changing signal may be misinterpreted as a change in the biology.

(d) Further problems arise with a number of extrinsic probes. After introducing the probes into the cell(s), they may gradually leak out of the cell again. This may even occur through active transport by the cell.

6. Solutions

(a) The most effective ways of avoiding artefacts due to loss of dye, movement, or bleaching involve using dyes which show a spectral shift in response to the variable that is measured, rather than simply showing an increase or decrease in fluorescence over the whole spectrum. These probes, first introduced by Tsien and his group (11), greatly simplify the process of calibration. If a dye shows a spectral shift which is dependent on a given variable, then the calibration becomes independent of concentration and will depend only on the relative position of the spectrum. Details of such calibrations are given below. Dyes that show a spectral shift are now the favoured probes for $[Ca^{2+}]_i$ and pH_i, and similar probes are available for $[Na^+]_i$ and $[K^+]_i$.

(b) Artefacts due to direct effects of or interactions with a drug may be suspected if a dual emission or excitation probe is used, and the changes in response are seen only at one wavelength or in the same direction at both wavelengths. If a probe is itself fluorescent, the control is simple—look at the 'response' in the absence of a cell, and be suspicious of responses that occur very fast (this will obviously depend on the method you use to apply drugs). It is worthwhile doing some simple experiments in a cuvette in a spectro-fluorimeter, if available, to check direct interactions between a drug and a dye. Obviously, if the change in signal is reproduced in a system without any cells present, the biological relevance may be doubtful—although such information has been used to some effect in interpreting biological responses.

(c) Bleaching can be minimized either by intermittent exposure of the preparation, if the phenomena studied are slow, or by reducing the light intensity with appropriate neutral density filters.

(d) To avoid or at least to detect movement or focusing artefacts, we have incorporated a video camera (see *Figure 1*) and a monitor into our system, so that we can always see what is happening to the cell, even though the microscope is housed within a light tight box. This means that artefacts due to changes in focus, to movement, etc are easily detectable, so long as one remembers to watch.

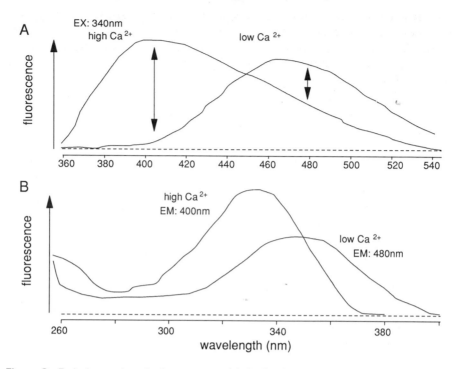

Figure 3. Emission and excitation spectra of Indo-1. (A) shows emission spectra in zero $[Ca^{2+}]$ (10 mM EGTA; 140 mM KCl; 10 mM Hepes; pH 7.2) and in high $[Ca^{2+}]$ (same solution, no EGTA, 1 mM $CaCl_2$). An excitation bandpass filter of 340 nm was used. Note the large change in fluorescence at 405 nm, with a rather smaller decrease in fluorescence at 480 nm as $[Ca^{2+}]$ rises. (B) shows excitation spectra under identical conditions. The spectrum shifts slightly to the right as $[Ca^{2+}]$ rises. If exciting light of 365 nm is used, there is minimal fluorescence change at 400 nm, with a significant change at 500 nm, 355 nm gives signals of equivalent amplitude, while 340 nm gives a larger signal at 400 nm and a small change at 480 nm.

7. Dyes with dual emission or excitation maxima

7.1 Dual emission probes

The principle of operation of a dye, Indo-1, that shows a spectral shift with the binding of Ca^{2+} is illustrated in *Figures 3–5*. *Figure 3A* shows emission spectra for 1 μM Indo-1 in a physiological saline in a cuvette, obtained using a Perkin Elmer spectrofluorimeter. Binding of Ca^{2+} to the Indo-1 shifts the peak of the excitation spectrum from about 480 nm towards about 405 nm. A rise in $[Ca^{2+}]$ leads to a decrease in fluorescence at 480 nm, a concurrent increase in fluorescence at 400 nm, and no change at about 450 nm. These spectra were obtained using an excitation wavelength of 340 nm. Note that this gives a large change in fluorescence (with changing $[Ca^{2+}]$) at 405 nm, but a rather small change at 480 nm. If longer excitation wavelengths are used (for example 365 nm), the spectra shift, so that the change at 400 becomes smaller, that at 480 more pronounced. This is illustrated in *Figure 3B*, in which excitation spectra are shown measured first at saturating $[Ca^{2+}]$, measured at 400 nm, and then in zero $[Ca^{2+}]$, measured at 480 nm. Signals of equivalent amplitude at the two wavelengths are obtained with an excitation at about 355 nm. We have chosen to use 340 nm, partly because we have a 340 nm filter, but more importantly, because it gives a large signal at 405 nm, which we can then sample alone to indicate changes in $[Ca^{2+}]_i$ when appropriate.

Movement of the cell, changes in focus, loss of dye from the cell will all cause a parallel loss of signal measured at both wavelengths. Thus, Indo-1 is optimally used by continuously measuring the fluorescence at both wavelengths, and by using the ratio of F_{405}/F_{480} as the primary signal monitored, as this should be determined solely by the concentration of $[Ca^{2+}]_i$. This is illustrated in *Figure 4*. I have chosen a record in which there is a small artefact on the original signals at both 405 and 480 nm (arrows). Notice that these are cancelled out in the ratio. Other dual emission probes currently available are detailed below.

7.2 Dual excitation probes

Some dyes show a shift in the excitation spectrum with a given variable. The best known is the Ca^{2+}-sensitive Fura-2. The emission wavelength measured is constant, at about 510 nm for Fura, while the excitation spectrum shifts with the binding of Ca^{2+}. The signal excited by 340 nm light increases with rising $[Ca^{2+}]_i$, while that excited at 380 nm decreases. The signal excited by 360 nm light should not change with changing $[Ca^{2+}]_i$. Another dual excitation probe widely used is BCECF to monitor intracellular pH, (pH_i). The need to switch the excitation wavelength to obtain a ratio (F_{340}/F_{380} for Fura-2) introduces several potential problems.

(a) The continual switching between wavelengths may take time, and reduces the temporal resolution of the ratio measurement. This is only important, of course, if the experiments require reasonably high temporal resolution of transients, for example. This has been partially overcome by using a spinning filter wheel to change the filters. Such systems are supplied with some commercially available packages for fluorescence microscopy (see below).

(b) The raw signal obtained with switching excitation light is almost uninterpretable as it stands, and has to be separated into the signals that derive from each excitation, either by a computer interface with appropriate software, or by analog circuits. This adds a level of complexity (and, therefore, a potential source of error) into the pathway. A number of commercially available packages will do these operations (see below). It is also relatively straightforward to build a circuit consisting of 'sample and hold' chips triggered at a given delay after switching a filter block or wheel, for example.

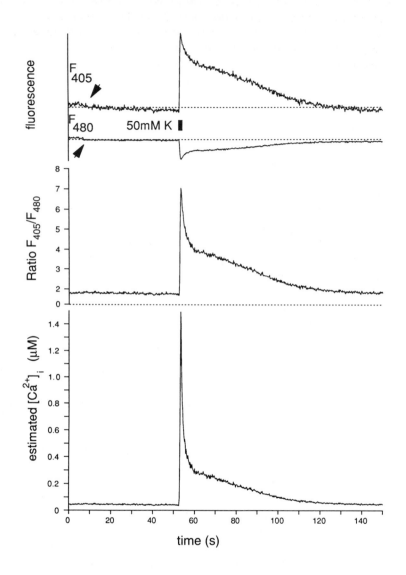

Figure 4. Records obtained from freshly dissociated mouse sensory neurons loaded with Indo-1 AM ester. An excitation wavelength of 340 nm was used. **Top** the raw fluorescence traces measured at 405 and 480 nm are shown. Exposure to a brief pulse of 50 mM K$^+$ to depolarize the cell caused an increase in fluorescence at 405 and a decrease at 480 nm. **Middle** shows the ratio of $(F_{405} - BG_{405})/F_{480} - BG_{480})$, where BG is the background signal. Note that the small artefacts at the start of the records (arrowed) are no longer apparent in the ratio. **Bottom** shows the change in $[Ca^{2+}]_i$ obtained from the ratio measurements (calculated and plotted using a spreadsheet). Note the change in the shape of the records following the calibration.

(c) Finally, the switching may generate significant vibration, a serious problem if simultaneous electrophysiological measurements are being made. This has been overcome in some laboratories by interposing a light guide between the switching unit and the microscope, so that transmission of vibration is minimized.

It is quite possible to gather data exciting most of the time at just one wavelength, switching manually to the other occasionally to ensure that there is limited loss of signal.

8. Fluorescence imaging

The objective of fluorescence imaging is to provide information about the spatial distribution of the fluorescence signals in a cell with time. Imaging techniques can also be used simply to follow changes in mean fluorescence from each of several cells within a field of view. Such studies have revealed spatial inhomogeneities in neurites, $[Ca^{2+}]_i$ hotspots in growth cones, waves of $[Ca^{2+}]_i$ in muscle cells, and interactions between cells. The basic requirements of the system are similar to those outlined above, except that a video camera (silicon intensified target, SIT; or charge coupled device, CCD) and image-processing hard and software replace the PMT measuring system. At present, most of those using such systems use relatively specialized and expensive hardware, but the arrival of ever more powerful microcomputers, perhaps coupled with optical disc storage capability, will make real time acquisition of such information more accessible.

The principle of imaging requires mapping of the fluorescence signal in relation to the cell, pixel by pixel. The fluorescence intensity registered at any one pixel will depend on the concentration of dye, the optical path length, and the variable to which the probe is sensitive, for example the $[Ca^{2+}]_i$. As the thickness of a cell will vary, the interpretation of such signals is complex. In most cases, the dual excitation probe Fura-2 has been used to measure $[Ca^{2+}]_i$. This means that for each pixel, the signal excited at 340 nm and that excited at 380 nm must each be measured and stored in the memory, and then the ratio calculated. The rate at which complete images can be acquired will depend on the rate at which the filter can be changed, the response properties of the camera, and the processing and storage capabilities of the computer. It is more complex to use a dual emission probe for imaging, as it requires the precise alignment of two cameras matching the image pixel for pixel to construct the ratio, although this has been achieved by Takamatsu and Weir (12).

The potential artefacts due to three-dimensional complexity are again much more extensive than a simple fluorimetric system. The technical requirements and problems have been extensively reviewed (13, 14). It seems inevitable that such systems will become simpler and more accessible in the near future, systems are currently commercially available from Joyce–Loebl and from SPEX (Section 13).

Imaging with con-focal scanning microscopy is still a difficult and expensive operation, but offers enormously enhanced spatial resolution. The principles of con-focal microscopy are outlined in reviews (15). Only laser-based con-focal systems can generate enough energy for ratio imaging in tissues. Those currently available use argon lasers with emission in the visible range, permitting the use of a non-ratio dye such as Fluo-3. There remain problems with the organization of UV imaging, as the chromatic aberration in the transmission of the UV by the optics results in differing planes of focus for the exciting and emitted light. There is considerable demand now for a ratio $[Ca^{2+}]_i$-sensitive dye that operates within the visible spectrum, and such a probe may well be developed soon.

9. Cellular autofluorescence

For most purposes, this is regarded as a nuisance, but it can convey interesting information about changes in cell metabolism. Autofluorescence excited by UV light of 340–360 nm derives mostly from mitochondrial NADH (which has an emission peak around 450 nm. As NAD is not fluorescent under these conditions, the signal reflects the state of the NADH/NAD ratio. The contribution from cytosolic pyridine nucleotides varies between cells, but is usually small.

Flavoprotein autofluorescence can also be monitored, having a maximal excitation peak around 450 nm, and emission at about 530 nm. Signals seem to be rather small. The responses are opposite in sign to those of NADH, as the oxidized form is fluorescent.

Cyanide (2 mM) and FCCP (1 μM) can be used to determine the peak and trough levels, increasing mitochondrial NADH fluorescence to a maximum and minimum, respectively.

10. Background fluorescence—definitions?

All fluorescence signals must always be identified with respect to a background signal. The true background is the cell without any dye, as this will include any autofluorescence and light scattering from the cell. In studies of single cells this is possible only if the dye is introduced as the free acid with a patch pipette, as the signal from the unloaded cell can be measured first. In AM (acetoxy-methoxy) dye-loaded cells, an average signal from unloaded cells can be used for cuvette experiments. For microfluorimetry, the signal from the dye is usually much brighter than the cells' autofluorescence (depending on the wavelengths used), so that it is not a major problem, and the background signal is taken simply as the window of the microscopic field without any cells. There are obviously potential errors if the autofluorescence is significant and changes with the experimental manipulations that are used, and perhaps this should be checked by measuring it directly. We found that with Indo-1 there is significant autofluorescence with excitation at 340 nm, but it changes similarly at both wavelengths used for measurements, so that even in response to cyanide or anoxia, which produces a maximal increase in autofluorescence, there was no significant change in ratio (6). With Fura-2, this may be more of a problem, as the autofluorescence generated by excitation at 340 nm will be significantly greater than at 380 nm, and the ratio of these two signals may change with a change in autofluorescence.

11. Introduction of probes into cells

Until recently, all $[Ca^{2+}]_i$-sensitive probes (Aequorin, Arsenazo III, Quin-2) had to be introduced into cells by injection with a microelectrode. In 1981, Roger Tsien (16) introduced a trick which involves esterifying the carboxylic acid moieties in several dyes with small hydrophobic groups. Acetoxy–methoxy groups have been most commonly used, resulting in so-called AM esters of many probes. The hydrophobic molecule should readily permeate cell membranes into the cytosol, (and other organelles) where non-specific esterases cleave the ester bonds to leave the charged dyes trapped in the cell. This has enormously increased the ease of doing such experiments, especially on small cells, and many commonly used probes are now available as AM esters.

11.1 Problems

(a) Some cells simply fail to take up the esters, especially cells from cold-blooded animals.

(b) The esters are hydrophobic, and tend to form micelles. These may then be taken up by phagocytosis, and accumulate in lysosomal vesicles, in which hydrolysis by lysosomal enzymes generates the free acid which then signals the high $[Ca^{2+}]$ acidic environment of the lysosomes. This problem may be overcome or reduced (i) by loading at 20 °C rather than at 35–37 °C, as the rate of endocytosis is very temperature sensitive, and (ii) by using amphiphilic molecules to help disperse the micelles. Such agents are Pluronic F-127 (200 mg/l), a detergent, and BSA at 0.01 to 0.1%. Pluronic is currently supplied by Molecular Probes with all acetoxymethyl esters along with a protocol for its use. Others have described accumulation of dyes in secretory granules, sarcoplasmic reticulum, and mitochondria. Most of these seem to be temperature sensitive, and this can be minimized by loading at room temperatures.

(c) In some preparations, loss of dye from the cells proceeds due to active transport by anion carriers. This may be prevented by probenecid (2.5 mM) or sulphinpyrazone, inhibitors of the transporter (17).

(d) Incomplete hydrolysis of the esters may lead to the formation of ionized derivatives which are not $[Ca^{2+}]$-sensitive but which are fluorescent. This will obviously increase the basal signal and reduce the changes, introducing errors into any attempt at calibration (see below). This effect may be minimized by incubating the cells at 35–37 °C after loading, and washing to promote complete hydrolysis of dye within the cell. Dye that is not hydrolysed should simply leak out of the cells again when they are washed. The AM ester of the $[Ca^{2+}]_i$ probe Fluo-3 is not fluorescent until it is hydrolysed, partially resolving this problem.

(e) Cleavage of the esters produces side products which may include acetic acid and formaldehyde. In practice, these are produced in such small quantities that they seem to cause little or no signs of damage. This should ideally be confirmed, however, wherever possible. For example, we have made electrophysiological recordings from neurons previously loaded with Indo-1, to confirm that the electrophysiological properties of the cells were unaltered by loading. It seems a good idea, nevertheless, to minimize loading time and concentration of dye and to wash the cells efficiently after loading.

Protocol 1: Loading cells with dye

1. Dissolve 1 mg of the AM (or any other AM ester) in 1 ml of acetone (this makes a 1 mM stock for Indo-1).
2. Divide into aliquots of 50 µl.
3. Evaporate off acetone in a freeze drier and store aliquots in a deep freeze.
4. When aliquots are to be used, resuspend in 50 µl of dimethyl sulphoxide (DMSO) (ideally anhydrous to reduce spontaneous hydrolysis of the dye). It may be helpful to use Pluronic F-127, in which case the dye is resuspended with DMSO made up with 20% Pluronic. Details for preparing this are supplied by Molecular probes with the Pluronic.
5. For use, soak cells in a dilution (1–2 µM) of the dye in a physiological saline at room temperature for about 30–45 minutes. The cells should then be washed and left for another 30 minutes to an hour or more at 35–37 °C to complete intracellular hydrolysis.

---

12. Specific probes

12.1 $[Ca^{2+}]_i$

Dual emission/excitation dyes

- *Fura-2*; dual excitation at 340 and 380 nm, measure fluorescence at 510 nm. Excitation by 360 nm light should not change at all (the isosbestic point) with changes in $[Ca^{2+}]_i$, but will change as dye is lost. K_D, 224 nM (125 mM K^+, 20 mM Na^+, 1 mM Mg^{2+}, pH 7.05, 37 °C).
- *Indo-1*; dual emission, excite at 340–365 nm, measure simultaneously at about 400 and 480 nm. Isosbestic point is about 450 nm, which will detect loss of

dye without a change in $[Ca^{2+}]_i$. K_D, 250 nM (125 mM K^+, 20 mM Na^+, 1 mM Mg^{2+}, pH 7.05, 37 °C).

Indo-1 seems to bleach more rapidly than Fura-2, but has also given rise to fewer problems with compartmentalization. Signals collected using Indo-1 also have a greater contamination with autofluorescence.

Single excitation/emission dyes

- *Fluo-3*, excited maximally at 490 nm, maximum emission at 525 nm. This has several useful features, although it cannot be ratioed. The long excitation wavelength limits photodynamic damage to the cells, and the signal/noise ratio is said to be very good owing to the large increase in signal with changing $[Ca^{2+}]_i$. K_D, 450 nM.

- *Rhod-2*; excitation at 556 nm, measure at 576 nm.

- *Quin-2*; excitation at 350 nm, measure at 490 nm. K_D, about 60 nM (in the presence of 1 mM Mg^{2+}).

12.1.1 Calibration

This has proved more than a significant problem in most experiments, so much so that you might consider whether it is really meaningful at all (see Section 12.1.2). Calibration of the fluorescence signal (F) requires the application of the equation:

$$[Ca^{2+}]_i = K_D \cdot (F - F_{min})/(F_{max} - F), \tag{1}$$

where F_{min} and F_{max} are the fluorescence signals with no calcium and with a concentration of calcium sufficient to saturate the signal, respectively, and with subtraction of background fluorescence. In addition, for dual emission or excitation dyes where there is a shift at both wavelengths used, all F terms are replaced by the ratio, R. The terms should be multiplied further by $(F_{\lambda2,\ min}/F_{\lambda,\ max})$ (see ref. 11) ($F_{\lambda2,\ min}$ and $_{max}$ are the minimum and maximum fluorescence intensities, respectively, measured at wavelength 2, the denominator of the ratio). For Indo-1, R is the ratio of fluorescence emitted at 400 nm to that emitted at 500 nm, and $F_{\lambda2}$ is the fluorescence emitted at 500 nm—a minimum ($F_{\lambda2,\ min}$ in saturating $[Ca^{2+}]_i$ and maximum ($F_{\lambda2,\ max}$) with zero $[Ca^{2+}]_i$.

The K_D values for many dyes are published under defined conditions, but it may be better for you to establish the effective K_D under appropriate conditions for your experiments (see *Protocol 2*).

The main requirement then is to establish F_{min} and F_{max} or R_{min} and R_{max}. If a non-ratio dye is used, then F is a function of the variable studied, the concentration of dye, and the properties of the system, and so F_{max} and F_{min} must be determined for every cell or every experiment (for spectrofluorimetry). If a ratio dye is used, R is independent of dye concentration. As the properties of the system change only slowly, R_{max} and R_{min} may be determined occasionally. How often should be assessed by seeing how variable the results are on successive days. In the best possible world, it is probably advisable to do this at the end of each day's work, as tomorrow the bulb may blow. In practice, few of us have the energy to do this after a long experiment, and it is done once a week or once a month. It must be redone if the system is significantly altered or if the bulb is changed.

While the K_D of the dye is an intrinsic property of the dye for given temperature, viscosity, pH, and ionic strength, the other parameters of the equation, depend critically on the light source, the transmission and light scattering of the system, and the preparation. These will, therefore, vary between systems and even between preparations. It is meaningless to use calibrations published or done elsewhere. The examples shown are chosen to illustrate points, not to be used to obtain values.

Protocol 2. Determination of K_D for a dye

To determine K_D for dye, use eqn 1.

To prepare known Ca^{2+}-EGTA mixtures in a physiological intracellular type saline to give known final $[Ca^{2+}]$ concentrations:

1. Prepare a stock of 100 mM Ca–EGTA. Make up solutions of 200 mM EGTA (use a pH meter, stir continuously and add 10 M NaOH to adjust the pH to 7.0 to get it into solution) and 200 mM $CaCl_2$. Using a pH meter, titrate the $CaCl_2$ into the EGTA. The end-point is the point at which the pH no longer changes following the additions.

2. Prepare two stock solutions (I and II), each containing (mM):

   • KCl or K glutamate, 140

   • NaCl, 10

   • Hepes, 7.5

   • Mg^{2+}, 1

   • MgATP, 2

   • 1 µM of the dye as the pentapotassium salt

   • pH adjusted to 7.2

3. Prepare solution I with addition of 10 mM EGTA, and solution II with 10 mM Ca–EGTA. (The concentration of EGTA must be much greater than that of the dye, so that the contribution of the dye to Ca^{2+}-buffering is insignificant.)

4. Solutions of predicted $[Ca^{2+}]$ are then made by mixing solutions I and II in proportions[a].

Equation 1 can then be used to obtain the K_D from the fluorescence ratios obtained.

But beware: the K_D is altered by temperature, viscosity, and ionic strength. The solution given in step 2 is an attempt to mimic the intracellular environment within reason.

---

[a]The exact proportions can be calculated from a variety of available computer programs (such as LigandY, available from P. Tatham or B. Gomperts, Dept of Physiology, UCL, Gower St, London WC1E 6BT). Some examples (for pH 7.2 at 25 °C) are:

pCa	Ca–EGTA	EGTA
6	6.86	1.14
6.5	5.25	2.75
7	3.01	4.99
8	0.46	7.54

(some details are also given in ref. 18).

Protocol 3. Calibrating fluorescence from cuvette experiments

There are numerous accounts of the various methods of calibrating fluorescence from cuvette experiments and the problems inherent in these (see, for example, ref. 19). One method follows:

1. At the end of the experiment, add EGTA and Tris base to give final concentrations of 4 mM and 30 mM, respectively, (raising the pH to >8.1 to increase the affinity of the EGTA for Ca^{2+}). This will reduce the fluorescence from dye that has leaked from the cells into the saline to a minimum. This assumes that $[Ca^{2+}]_i$ is not reduced by the fall in $[Ca^{2+}]_o$[a].

2. Lyse the cells with a detergent: digitonin (25 µM) or Triton-X (0.1%). This frees the dye into the saline which now has low $[Ca^{2+}]$ to give R_{min}.

3. Add excess (>4 mM) $CaCl_2$ to determine R_{max}.

It may be possible to obtain R_{max} without lysing the cells using a non-fluorescent Ca^{2+} ionophore such as ionomycin or 4-bromo-A23187 (5 µM) in place of the detergent in this sequence. Ionomycin operates most efficiently at alkaline pH (ideally around 9, but satisfactorily at 7.6–8.0).

---

[a]An alternative, (see ref. 19): add 50–100 µM $MnCl_2$ first. This quenches the fluorescence from any dye that has leaked from the cells. Addition of the heavy metal chelator diethylenetriaminepentaacetic acid (DTPA) (200 µM) then removes the Mn^{2+}. Lysis of the cells with the detergent then liberates the dye into the cuvette, where R_{min} and R_{max} can be determined as indicated above. From these signals is subtracted the fluorescence signal due to the extracellular dye.

Protocol 4. Calibration for microfluorimetry *'in vitro'*

This is a straightforward procedure that may yield the best approximation, but will probably be incorrect, as the properties of the dye change slightly within the cytosolic environment.

1. Prepare a saline to resemble the intracellular milieu as closely as reasonable. We have used (mM):
   - KCl or K glutamate, 140
   - NaCl, 5
   - Hepes, 7.5
   - Mg^{2+}, 1
   - MgATP, 2
   - pH adjusted to 7.2
   - Into 1 aliquot, add 1 mM $CaCl_2$, to another, 1 mM EGTA.

2. Make small wells that can fit on the microscope stage—we use small Perspex rings of about 5 mm diameter, which adhere to a coverslip using a thin layer of silicon vacuum grease. These must all be very clean.

3. Place equal volumes of each saline (about 200 µl) in each of two wells on the coverslip and record the fluorescence signal, which represents the background.

4. Taking care not to contaminate either solution, add Indo-1 or Fura-2 pentapotassium salt to a final concentration of 1 or 2 µM. Mix, and record the fluorescence signals again to give values for R_{max} and R_{min}.

5. The properties of several dyes are altered by the intracellular viscosity, which tends to give an underestimate of $[Ca^{2+}]_i$ in experiments with Fura-2, as the fluorescence excited at shorter wavelengths is enhanced. This effect has been compensated for by adding Ca^{2+}-free gelatin at about 4% or sucrose (2 M) to raise the viscosity of the saline, the problem is discussed at some length in ref. 20.

For microfluorimetry *in situ*, if the cells are to be studied with patch clamp techniques it should be possible to control $[Ca^{2+}]_i$ sufficiently to determine R_{min} and R_{max} directly. Use 0.1 mM of the pentapotassium salt of the dye. It is also possible to carry out this procedure for AM-loaded cells.

I have tried bathing cells in a $[Ca^{2+}]_o$-free saline (with 1 mM EGTA), making whole cell recordings with a pipette filling solution containing 10 mM BAPTA to obtain R_{min}.

R_{max} was obtained with a $[Ca^{2+}]_o$-containing saline, making whole cell recording without added $[Ca^{2+}]_i$ buffers and effectively electropermeabilizing the cell by voltage clamping the membrane potential to $-200\,mV$, which causes (rather slow) dielectric membrane breakdown, or even pulling off a patch, which is often sufficient to kill the cell. Both lead to a rise in $[Ca^{2+}]_i$ followed by the slower loss of dye from the cell (see also refs 21, 22).

In some preparations, $[Ca^{2+}]_i$ ionophores are used to raise $[Ca^{2+}]_i$ maximally, but in our hands, $[Ca^{2+}]_i$ fails to rise as much with ionomycin or 4-bromo-A23187, even at high concentrations, as it does simply with depolarization.

12.1.2 To calibrate $[Ca^{2+}]_i$?

There are problems with all forms of calibration. These are discussed extensively in a number of papers (6, 10, 13, 17–23). Perhaps it is worth seriously questioning whether you want to put absolute numbers next to your traces, and whether they are real. A number of investigators have accepted that their calibrations are not really meaningful, and present data simply as a ratio. If you really need to know the true values of $[Ca^{2+}]_i$ for your system, you will probably have to go to some trouble to make sure the determinations are meaningful.

A further consideration is the non-linearity of the ratio/$[Ca^{2+}]_i$ relation (*Figure 5A*). This means that if ratios are converted to $[Ca^{2+}]_i$, the shapes of the traces may be significantly altered (see *Figure 4* and *Figure 5B,C*), becoming more significant the higher the $[Ca^{2+}]_i$ (or the ratio). Measurements of rates of change, fitting of exponentials to traces, and other similar manipulations will be quite different if you use ratios or if the traces are calibrated, and will always be of rather doubtful value except for comparative purposes.

The rates of change of signals may even be limited by the response properties of the probe—the off rate of some probes may have time constants of the order of 50 ms. The relatively high affinity of Indo-1 or Fura-2 for Ca^{2+} also means that these dyes may be saturated at around 1–$2\,\mu M$ $[Ca^{2+}]_i$, so that any further increase in $[Ca^{2+}]_i$ fails to raise the ratio significantly (see *Figure 5A*). For this reason, a dye like Fluo-3, which has a lower affinity for Ca^{2+} (K_D about 450 nM) may be more useful if the processes studied are operating at high $[Ca^{2+}]_i$ levels.

12.2 pH_i

There is a choice of several such probes. The choice should be made on the pK_a of the probe, in that it should ideally lie within the physiological range expected.

Dual excitation dyes:
- BCECF; excite at 440 nm and 500 nm, measure at 510–550 nm. pK_a near 7.0. An acidification decreases fluorescence excited by 490–500 nm while that excited at 440 nm remains largely unchanged. The ratio of 440/500 nm fluorescence is used to monitor pH_i.

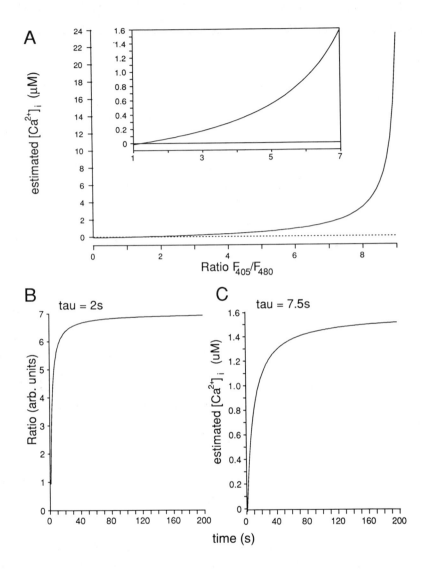

Figure 5. A calibration curve of ratio (F_{405}/F_{480}) vs $[Ca^{2+}]_i$ for Indo-1 in our system. **(A)** shows a full curve, but note that above about 4 μM, changes in ratio for a large change in $[Ca^{2+}]_i$ are not resolved. Inset is the range over which most physiologically relevant changes occur, and the non-linearity is still noticeable. In **(B)**, a simple exponential increase in ratio is plotted. This was simply calculated in a spreadsheet, and had a time constant, tau, of 2 seconds. **(C)** shows the result when this change in ratio is calibrated according to the curve in (A). The non-linearity of the calibration translates into a change in $[Ca^{2+}]_i$ which is fitted by an exponential for which tau is 7.5 seconds.

256

- SNAFL-1; excite at 470–530 nm and 550–560 nm; measure at 600 nm; pK_a 7.85. SNAFL-1 can also be used as a dual-emission dye (see below).

- SNARF-6; excite at 488–530 nm and 560 nm, measure at 610 nm; pK_a, 7.75.

Dual emission dyes:

- DCH2 (dicyanohydroquinone, also known as DHPN (dihydroxyphthalonitrile); excite at 405 nm; measure at 435 and 480 nm. Acidification increases fluorescence at 435 nm and decreases that at 480 nm. pK_a is about 8.0. DCH2 leaks out of cells, as it is not strongly charged, but as the ester is largely non-fluorescent, it can be used in the continuous presence of the ester.

- SNAFL-1; excite at 510–540 nm; measure at 540–550 and 620–650 nm. Acidification increases fluorescence at around 540 nm, with little change or a slight decrease at 620 nm. pK_a is 7.85.

- SNARF-2; excite at 488–530 nm; measure at 580 and 640 nm. Acidification increases fluorescence at 580 nm and decreases it at 640 nm. The pK_a is about 7.5.

Single wavelength measurements

- Carboxyfluorescein; excite at 480 nm and measure at 520 nm. However, this is poorly retained by cells, and inserts into organelles as well as the cytosol. As lysosomes and secretory granules have an acidic and mitochondria an alkaline pH, these may distort the measurements depending on their density.

Calibration of intracellular pH is much easier than it is for $[Ca^{2+}]_i$, as the K–H ion exchanger nigericin is reasonably effective at equalizing intracellular and extracellular pH, (*Protocol 5*).

Protocol 5. Determination of intracellular pH

The pH_i can be calculated from signals with BCECF using equation 2 (ref. 23):

$$pH_i = pK_a + \log (R - R_{min})/(R_{max} - R) + \log(F_{430,\ min}/F_{430,\ max}). \qquad (2)$$

1. Bathe cells in a saline containing isotonic replacement of Na^+ with K^+. Omit added Ca^{2+} if the cells are excitable, as this saline will depolarize the cells and raise $[Ca^{2+}]_i$ which may have secondary effects on proton distribution.

2. Prepare similar salines with pH adjusted to a minimum of 3 different levels—such as 5.5, 7.0, and 8.5.

3. Add 10 µM nigericin to each saline, and expose the cells sequentially to each. A plot of ratio against pH of the bathing saline under these conditions should give a reasonable calibration (see ref. 23).

12.3 $[\text{Na}^+]_i$

Intracellular sodium can be measured using either the dual excitation dye SBFI or the dual emission Fcryp-2.

- SBFI is excited at 340 and 380 nm and fluorescence is measured at 505 nm—very similar to the requirements for Fura-2. However, the spectrum appears to shift *in vivo*, such that there is little change at 340 nm, while the signal decreases at 380 nm with an increase in $[\text{Na}^+]_i$.

- Fcryp-2 is excited at 340–355 nm and measured at 405 and 480 nm—very similar to the requirements for Indo-1.

Calibration should be relatively straightforward, using the ionophorous antibiotic nystatin (50 µg/ml) to equilibrate Na^+ and K^+ across the cell membrane. Simply varying the ratios of Na^+/K^+ in a physiological saline should change the $[\text{Na}^+]_i$ predictably to give ratios representative of a range of Na^+ concentrations that allow construction of a calibration curve.

12.4 $[\text{Mg}^{2+}]_i$

There are magnesium-selective versions of both Indo-1 and Fura-2, with optical properties similar to those of the parent dye. There are some problems with these, in that they remain sufficiently sensitive to large changes in $[\text{Ca}^{2+}]_i$ (i.e. into the micromolar range) so that care must be taken in the interpretation of the signals obtained. The ionophore bromo-A23187 will carry Mg^{2+} across the membrane as well as Ca^{2+}, so that it may prove possible to obtain R_{min} and R_{max} by bathing cells in a Ca^{2+}- and Mg^{2+}-free saline, adding bromo-A23187 (1 µM??) and then adding Mg^{2+} until the ratio is saturated (which may have to be >10 mM, as the K_D for Mg^{2+} of mag–Indo is about 2.7 mM, and that of mag–Fura, about 1.5 mM).

12.5 $[\text{Cl}^-]_i$

SPQ: excite at 350 nm, measure at 440 nm. Considering the importance of Cl^- conductances in GABA and glycine-mediated inhibition, this may prove a most useful tool, but it has not been extensively used. The principle of operation is rather different to most of the indicators mentioned here, as binding with Cl^- quenches rather than increasing fluorescence. The dye is also not as sensitive inside cells as it is alone, and tends to leak rapidly from cells after loading. These factors have limited its usefulness, (but see ref. 24).

Calibration can be achieved using a Cl^-/OH^- exchanger, tributylin, together with nigericin, a K^+/H^+ exchanger, thus clamping $[\text{Cl}^-]_i$ by varying pH.

12.6 Mitochondrial membrane potential ($\Delta\psi_m$)

Use Rhodamine 123. Load with 10 µg/ml for 10–15 minutes and then wash well. Excite at 490–500 nm, emission peak at 530 nm. An increase in fluorescence indicates mitochondrial depolarization, and is readily tested with the uncoupler carbonylcyanide p-fluorophenoxymethyl hydrazone (FCCP) 1–2 µM. I am not aware of any serious attempts to calibrate the signal in cells. In isolated mitochondria, it can be calibrated using the K^+ ionophore valinomycin and imposing various K^+ potentials (see ref. 25).

12.7 $[Zn^{2+}]_i$

As zinc is accumulated in some neurons (for example hippocampal Mossy fibres) and has been shown to modulate both excitatory and inhibitory events in the CNS, such a probe may be of some interest. A recent report suggested that N-6-methoxy-8-quinolyl-p-toluenesulphonamide (TSQ) may be useful for following changes in Zn^{2+} content of boutons.

13. Conclusions

It should be clear that this technology is both powerful and wideranging. Once established in a laboratory, it is likely to be continuously developed as new probes and new questions develop. Imaging and con-focal imaging technologies are becoming more practical propositions for more labs. At present, they are expensive options, and photometric microfluorimetry seems to me to be the best alternative. I would advocate building a system that leaves options for both dual emission and dual excitation probes. The cheapest way to do this is to build your own additions to interface the PMTs with a suitable fluorescence microscope. Bundled packages that include computer software are currently available from Newcastle Photometric Systems (Newcastle upon Tyne, UK); Joyce–Loebl Ltd (Gateshead, UK) and SPEX (Stanmore, UK), in increasing order of expense. The approach of each of these is rather different, and I can pass no direct judgment on them, as I have seen only their literature and demonstration discs, which are readily available. The software will doubtless improve with time.

My hope in writing this chapter is that you might avoid some of the practical pitfalls which we had to discover by our own errors.

Acknowledgements

I thank the Wellcome Trust and the Royal Society for financial support. I also thank Tim Biscoe for his helpful advice on the text and Duncan Farquharson for his help in building our set-up.

References

1. Hill, D. K. and Keynes, R. D. (1949). *J. Physiol.* **111**, 278.
2. Ashley, C. C. and Ridgeway, E. B. (1970). *J. Physiol.*, **209**, 105.
3. Chance, B. and Williams, G. R. (1955). *J. Biol. Chem.*, **217**, 395.

4. Tsien, R. Y. (1989). *Ann. Rev. Neurosci.*, **12**, 227.
5. Nachshen, D. A. (1985). *J. Physiol.*, **363**, 87.
6. Sackmann, B., Edwards, F., Konnerth, A., and Takahashi, T. (1989). *Q. J. Physiol.*, **74**, 1089.
7. Regehr, W. G., Connor, J. A., and Tank, D. W. (1989). *Nature*, 341, 533.
8. Biscoe, T. J. and Duchen, M. R. (1990). *J. Physiol.*, **428**, 31.
9. Lakowicz, J. R. (1983). *Principles of fluorescence microscopy*. Plenum Press, NY.
10. Tsien, R. Y., Rink, T. J., and Poenie, M. (1985). *Cell Calcium*, **6**, 145.
11. Grynkiewicz, G., Poenie, M., and Tsien, R. Y. (1985). *J. Biol. Chem.*, **260**, 3440.
12. Takamatsu, T. and Weir, W. G. (1990). *Cell Calcium*, **1**, 111.
13. Bolsover, S. R., Silver, R. A., and Whittaker, M. (1991). In *Electronic light microscopy*, (ed. D. Schotton). Alan R. Liss, NY (in press).
14. Tsien, R. Y. and Harootunian, A. T. (1990). *Cell Calcium*, **11**, 93.
15. Fine, A., Amos, W. B., Durbin, R. M., and McNaughton, P. A. (1988). *Trends Neurosci.*, **11**, 346.
16. Tsien, R. Y. (1981). *Nature*, **290**, 527.
17. DiVirgilio, F., Steinberg, T. H., Swanson, J. A., and Silverstein, S. C. (1988). *J. Immunol.*, **140**, 915.
18. Williams, D. A. and Fay, F. S. (1990). *Cell Calcium*, **11**, 75.
19. Rink, T. J. and Pozzan, T. (1985). *Cell Calcium*, **6**, 133.
20. Poenie, M. (1990). *Cell Calcium*, **11**, 85.
21. Almers, W. and Neher, E. (1985). *FEBS Lett.*, **192**, 13.
22. Benham, C. D. (1989). *J. Physiol.*, **415**, 143.
23. Eisner, D. A., Nichols, C. G., O'Neill, S. C., Smith, G. L., and Valdeolmillos, M. (1989). *J. Physiol.*, **411**, 393.
24. Krapf, R., Berry, C. A., and Verkmann, A. S. (1988). *Biophys. J.*, **53**, 955.
25. Emaus, R. K., Grunwald, R., and Lemasters, J. J. (1986). *Biochim. Biophys. Acta*, **850**, 436.

10

Ion-selective electrodes

EVA SYKOVÁ

1. Introduction

The invention of ion-selective microelectrodes (ISMs) has opened up new possibilities for experimental physiology. The measurement of the activity of specific types of ions in different cells and in the extracellular space, together with the resting potential, action potential, and extracellular potential shifts, has increased our knowledge about the relationship of ion activities and membrane transport mechanisms in excitable tissues. Measurement with ISMs is the only available method which provides simultaneous information about ion activity and electrical parameters of a cell in living tissue. Therefore, the measurement of ion activities with ISMs has become, in the last 20 years, an important tool in microelectrode studies not only in neurophysiology, sensory physiology, muscle physiology, but also in epithelial studies and in renal physiology.

1.1 Basic theory of ion-selective microelectrodes

ISMs are miniaturized potentiometric sensors whose main component part is an electrochemical membrane. The membrane separates two solutions of different ionic concentration and is selectively permeable (ideally) to a single ion. When there is a concentration gradient between the two solutions the ion of interest passes across the membrane (carrying its charge with it). This generates a potential difference that, in turn, impedes further movement and gives a steady-state response. The potential generated is determined by the concentration difference between the two solutions and forms the basis of the measurement. When the ion concentration of the solution on one side of the membrane is known, the potential generated across the membrane can (with appropriate calibration) be used to determine ion concentration in an unknown solution. In ISMs, the selectively permeable membrane occupies the tip of a micropipette which is then filled with a solution of known ionic composition. The tip is then introduced into the tissue or sample solution and the resultant potential measured and calibrated. Any change in the ion concentration due to experimental manipulations manifests itself as a change in potential.

1.2 Solid and liquid membrane ISMs

From the historical point of view we can divide ISMs into two categories, those with fixed-site or solid membranes and those with liquid membranes. The first ISMs were made of ion-selective glass. Their disadvantage was their relatively low selectivity and the large size of the active surface. Today, in electro-physiological studies, the trend is clearly towards using liquid-membrane micro-electrodes. Microelectrodes based on liquid membranes are the best sensors for measuring dynamic changes in the activity of biologically important ions *in vivo*. They have become widely used because of their relatively easy construction and miniaturization. This chapter will describe how to make ISMs with liquid membranes and how to use them in living tissue. More detailed theory and some alternative methods of their preparation can be found in several monographs (1–6).

2. Ion carriers

Each ISM consists basically of a liquid membrane in the tip of a glass micropipette, the main component of which is an *ionophore* (ion-bearer, ion-carrying agent, carrier). The ionophore causes selective ion permeability in artificial membranes, due either to ion-specific channels or ion-carriers. However, only ion-carriers have become widely used as the ion-selective components of liquid membranes of current ISMs because of their superior properties (selectivity, transport rate, thickness of membrane, etc.). The ion-carriers used today can be divided according to their selectivity as:

● cation-selective carriers
● anion-selective carriers

We can also subdivide them as:

● electrically neutral carriers
● electrically charged carriers

Most cation-carriers are synthetic compounds, either antibiotics (for example the K^+-selective carrier valinomycin), crown compounds, cryptans and non-macrocyclic compounds, for example lipophilic oxa-amides. At this time the non-macrocyclic molecules are the most widely used neutral carriers (for further details see ref. 6).

The most important properties of the carriers used in ISMs, are:

● induction of selective permeability of the liquid membrane (i.e. the exclusion of membrane permeability for other ions)
● optimal ion-exchange rate (i.e. sufficiently fast response time, usually in the range of milliseconds to a few seconds).
● lipophilicity (carriers with low lipophilicity will be rapidly lost from the electrode membrane into the surrounding solution resulting in a carrier-depleted membrane surface)

Neutral carriers, (generally the most selective class), are, for analytical applications, incorporated into liquid membranes to form an organic phase separating the sample solution containing the given ion from the aqueous solution filling the microelectrode (see *Figure 1*). The neutral carriers are able to transport the respective ions from the sample solution, across the membrane to another aqueous solution. Liquid membranes in ion-selective microelectrodes consist of:

● the carrier

● the membrane solvent (plasticizer), for example nitrobenzene

● the membrane matrix (polymer), for example PVC

● if necessary, other membrane additives.

Many ion-exchange membranes can be purchased essentially as complete 'cocktails' that include solvent, polymer, and such other additives as necessary. These are ready for use without further treatment and offer an easy start to the beginner. Fluka Chemical Corp. make cocktails for the detection of Ca^{2+}, H^+, K^+, Na^+, Mg^{2+} and Li^+.

Optimal liquid membrane properties are:

● chemical stability

● inertness

● adequate viscosity

● adequate dielectric constant

● high lipophilicity

● low toxicity

● low electrical resistance of membrane

● biocompatibility.

As mentioned above, the ion-exchanger membrane separates two ionic solutions of different concentrations. When an ISM is introduced into a tissue, or calibration solution, the potential which develops across the ion-exchanger membrane depends on the activity of the given ion. We speak of selectivity because, although the membrane potential can be influenced by various types of ions, from them the ion-exchanger membrane 'selects', i.e. gives priority to, a particular type. At present there are sensors (ion-exchangers) for the measurement of H^+, Li^+, Na^+, K^+, Mg^{2+}, Ca^{2+}, Cl^- and HCO_3^-. Details of selectivity factors, activity coefficients, liquid-junction potentials, response time, and other theoretical aspects of ion-selective microelectrodes can be found in various monographs (1–6) and in the catalogue of *Ionophores for ion-selective electrodes* published by Fluka Chemical Corp.

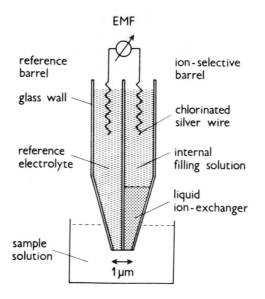

Figure 1. Schematic cross-section of a double-barrelled ion-selective microelectrode.

3. Steps in the construction of ion-selective microelectrodes with liquid membranes

There follows a step-by-step description of how to construct ISMs. Although some ion-selective electrodes with clinical application are available commercially, the best way to obtain microelectrodes for basic physiological research is to make them oneself in the laboratory.

Figure 2 shows, schematically, the major classes of microelectrodes:

- a single-channel ISM
- a two-channel ISM made either from theta glass-tubing or from two pieces of twisted glass tubing
- a concentric two-channel ISM
- a side-pore ISM

Nowadays, two-channel microelectrodes (double-barrelled ISMs) are principally used, especially in excitable tissues. Single channel ISMs are unsuitable for measurements in excitable tissues, since, together with the potential due to changes in relevant ionic activity, the microelectrode also picks up biopotentials (such as action potentials, field potentials, and synaptic potentials). When recorded simultaneously, the biopotentials corrupt the measurement of ion activity changes. Biopotential interference can, however, be abolished if the reference electrode is in the immediate vicinity of the tip

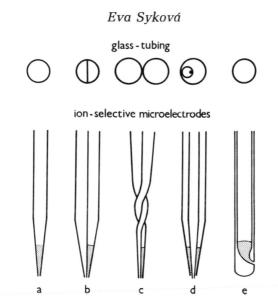

Figure 2. Different types of glass tubing (cross-sections) and longitudinal section of ion-selective microelectrodes pulled from the tubings. (a) single-barrel microelectrode; (b) double-barrelled microelectrode pulled from theta glass; (c) double-barrelled microelectrodes pulled from two twisted glass tubes; (d) concentric microelectrode with inner glass filament in the reference barrel; (e) side-pore electrode.

of the ISM, a requirement best fulfilled with *double-barrelled microelectrodes*. One barrel is filled with saline, or some other indifferent solution, and acts as the reference electrode, while the other is the actual ion-selective electrode. It is important to stress that care must be exercised in coating the reference silver (Ag) wire with silver chloride (AgCl). An inadequate coat will mean that the electrode potential drifts, thus obscuring any biological changes.

The construction of ISMs does not differ substantially with the type of ion-carrier (liquid ion-exchanger) used, the differences being only in the internal filling solution of the ion-selective barrel, in calibration, and in storing microelectrodes. The construction of a double-barrelled microelectrode involves the following steps:

- cleaning glass tubes chemically or with ultrasound (Section 4.2)
- pulling micropipettes to a tip diameter of 0.5–1.0 μm (*Protocol 1*)
- silanizing the ion-selective barrel tip to achieve a hydrophobic film on the inner walls (*Protocol 2* and Section 5)
- filling the tip of the ion-selective barrel with liquid ion-exchanger and a corresponding internal filling solution (*Protocol 3* and Section 6)
- measuring the resistance of each barrel
- calibrating, storing, and using ISMs

4. Glass tubing and pulling of micropipettes

ISMs can be pulled from various types of glass tubing. The important parameters of glass to consider are its:

- softening point

- water resistivity

- tendency to form microcracks in the glass wall near the tip during the pulling procedure

4.1 Types of glass

Borosilicate glass (Pyrex) seems to be best, especially for intracellular use, i.e. for ISMs with a fine tip diameter (7). The aluminosilicate glass is highly water resistant (8): hydration of the glass can alter the tip geometry and, therefore, the tip potentials. Different shapes of glass tubing are available commercially but only some are suitable for preparing ISMs (for example Clark Electromedical, Pangbourne, UK). The most frequently used tube shapes are shown in *Figure 2*. Key features are the thickness of the wall and diameter of the tubing. For intracellular recordings which require ultrafine tips, the ideal diameter of the tubing is about 1 mm, with a thin glass wall. However, where possible (for example larger cells, extracellular recordings), the diameter of the tube should be greater, 2.0–2.5 mm, and the walls correspondingly thicker. Greater diameter tubing allows easier manipulation of the electrodes during filling and removal of air bubbles. A thick wall, or septum in theta glass, is recommended to avoid an electrical connection between the reference barrel and ion-selective barrel in double-barrelled microelectrodes. Tubing with inner filaments (much easier to fill) is often used in the reference barrel, but is not suitable for the ion-selective barrel, since the filaments can cause improper silanization of the tip and difficulties in the removal of air bubbles between the ion-exchanger and internal filling solution.

4.2 Cleaning the glass

An important step is cleaning the glass tubes before use. The purpose is to get rid of dust particles and grease to improve reactivity during silanizing and removal of air-bubbles. Some commercially available tubes are clean enough to be used without any treatment, although others need careful cleaning with strong acids or organic solvents, for example by soaking the tubing in concentrated H_2SO_4 and H_2O_2 (3:1) for at least 2 hours. Any such cleaning procedure requires careful subsequent washing of the tubes with distilled or double-distilled water. It is useful to clean several dozen tubes simultaneously and then wash them not less than 10 times with distilled water. The tubing can also be cleaned with ultrasound.

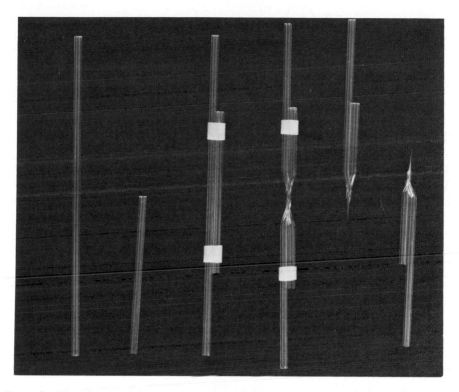

Figure 3. Procedure for the preparation of double-barrelled micropipettes. Left to right: two glass tubes of different lengths, fixing the tubes together by a polythene ring, twisting and slight (about 3 mm) pulling of tubes with a Bunsen burner, two micropipettes after subsequent pulling in a microelectrode puller.

4.3 Double-barrelled ISMs

For double-barrelled microelectrodes used for extracellular recordings, two twisted tubes are frequently used (*Figure 2c*). The preparation of twisted tubings is shown in *Figure 3*. Essentially, two tubes are fixed together, twisted through 360° and pulled (*Protocol 1*). These micropipettes have several advantages, i.e: the diameter of the tubing can be as large as 2.5 mm with relatively thick walls; the different lengths of the tubes helps to prevent mixing of the solutes from both barrels. The septum between the barrels does not crack and, even in the tip, the septum is thicker than in theta glass (*Figure 2b*), therefore preventing communication between the barrels as well as making filling of the reference barrel with ion-exchanger easier.

Coaxial, low resistance ISMs (see *Figure 2d*) have been described by Ujec *et al.* (9). In these ISMs, electrical parameters were shown to be improved (time constant, noise level, total resistance). The preparation of these ISMs is, however, considerably more difficult and, therefore, they are less widely used.

The electrodes are pulled from tubing into which another, finer, tube with an inner filament is introduced. The external tubing is filled with ion-exchanger, while the internal tube serves as the reference channel (9).

4.4 Single-barrelled microelectrodes

Side-pore electrodes (see *Figure 2e*), designed by Vyskočil and Kříž (10) for measurements during intensive muscle work, are pulled from conventional glass tubing. The tip of the microcapillary tube is sharpened on a fine grinding wheel to obtain a hypodermic-needle-shaped tip of 30–100 μm in diameter. When this tip is quickly (in 0.5–1.0 s) passed through a small gas flame (use a fine Bunsen burner), a side-pore channel of 3–5 μm is obtained. This micropipette is then filled with ion-exchanger as with any other single channel micropipette.

4.5 Electrode pulling

For pulling micropipettes one can use any commercially available or home-built puller or a more sophisticated one (for example Brown–Flaming type) for intracellular experiments. The requirements are the same as for conventional microelectrodes (non-selective ones), i.e. tip diameter, length of the shank (the pulled part of the tubing), and the shape of the tip (see Chapters 1 and 2). Micropipettes for intracellular ISMs can be bevelled in the same way as any conventional microelectrodes. *Protocol 1* describes the pulling of a double-barrelled microelectrode.

Protocol 1. Fabrication of double-barrelled microelectrodes

1. Take two glass tubes (up to 2.5 mm diameter), one shorter than the other (for example 6 and 10 cm lengths): see *Figure 3*.

2. Fix the tubes together either with a polythene ring (cut from a polythene tube) or with Heatshrink.

3. Heat the middle of the tubing with a small Bunsen burner until the glass softens.

4. Twist the tubing manually through 360° and pull the two sections approximately 3 mm apart. Allow to cool.

5. If desired, remove the Heatshrink or polythene tubing.

6. Fix *one end* of the twisted tubing in the upper chuck of a vertical microelectrode puller. Do not attempt to fix the other end into the lower chuck (tubing twisted in a Bunsen flame by hand is not straight enough to be fixed at both ends without cracking).

7. Preheat the tubes gently until the thinnest part starts to soften and then swiftly attach the tube to the lower chuck and pull apart.

NB: A tip diameter of 0.5–1.0 μm is ideal and this necessitates careful adjustment of heat and pull settings. These are often peculiar to individual models of puller and thus trial and error is necessary to achieve good results.

---

5. Silanization

The surface of glass is quite hydrophilic and therefore repels organic liquids. Silanized glass becomes hydrophobic. Therefore the silanization, first introduced for ISMs by Walker (11), dramatically improves the lifetime of ISMs. It also increases electrical resistivity of the glass surface and facilitates the uptake of the organic membrane solution (ion-exchanger).

Today, various techniques to silanize the ion-selective barrel are used, from the simple to the very sophisticated. The methods differ according to whether single- or double-barrelled ISMs, ISMs with fine tips to be used intracellularly, or ISMs with tips 2–4 µm for extracellular recordings are to be made. In principle, ISMs are dipped into an organic solution containing a reactive silicon compound (for example chlorosilanes, aminosilanes, siloxanes (Fluka Chemical Corp.)) dissolved in carbon tetrachloride or xylene. Alternatively the silanization technique can be based on treatment with a vapour phase of reactive silicon compounds.

5.1 Dip silanization

The simplest method is to dip the microelectrode into a solution of 1–5% of the reactive silicon compound, the smaller the tip, the less concentrated silicon solution should be used (10–12).

In the case of double-barrelled microelectrodes, the reference barrel can be protected either by injecting it with a drop of distilled water which fills the tip by capillarity, thus preventing it sucking in silicon, or by pressure applied to the reference barrel. When dipped into the silicon solution, a column of about 200–500 µm is drawn up into the ion-selective barrel. It can be either expelled from the shank of the micropipette several times or the micropipettes can be placed in an oven and heated to 150–200 °C for about 1 hour. After evaporation of water and silicon solution the micropipettes are ready for filling.

5.2 Vapour silanization

The more sophisticated, but also more time-consuming, are silanization techniques based on treatment with silicon vapour, these are necessary especially for silanizing 'intracellular ISMs' with smaller tips. The simplest method is described in *Protocol 2* and *Figure 4*.

Protocol 2. Silanization of intracellular microelectrodes

1. Place several microelectrodes in a holder on a glass plate.
2. Cover with a glass beaker.
3. Preheat the electrodes and glass beaker in oven (about 30 min, 150 °C).
4. Inject a small drop (10 µl) of the pure silanization reagent, by syringe, into the beaker. The silicon compound (for example dimethylamino

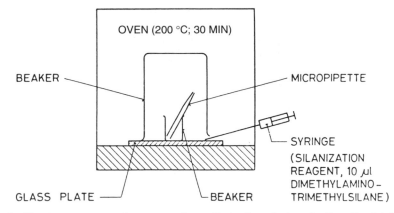

Figure 4. Simple arrangement for vapour-phase silanization of micropipettes. For details see the text. (From ref. 6 with permission.)

Protocol 2. *Continued*

    trimethylsilane) immediately evaporates. Electrodes should be kept in contact with the silicon vapour for about 30 min at 200 °C (13).

5. Remove and dry the electrodes at 150–200 °C for one hour.

6. Allow to cool.

This method, shown in *Figure 4*, is relatively simple, but can be used only for single-barrel microelectrodes.

5.2.1 Vapour silanization of double-barrelled microelectrodes

For double-barrelled microelectrodes, vapour silanization requires a special apparatus for the selective vapour-phase silanization of the ion-selective barrel alone. An example is shown in *Figure 5* (14). In principle, a stream of gas (for example nitrogen) carries the silicon vapour through the ion-selective barrel. The reference barrel is protected by a stream of nitrogen. Another silanization set-up used for double-barrelled microelectrodes made from theta glass is shown and described in *Figure 6* (15).

    In summary, proper silanization is an important step in the fabrication of ion-selective microelectrodes. In almost every laboratory, some improvement of the above methods is used for silanizing micropipettes.

6. Filling the micropipettes

The methods of filling micropipettes can be divided into two categories:

- front-filling (i.e. filling through the tip)
- back-filling (i.e. filling via the unpulled end of glass tubing

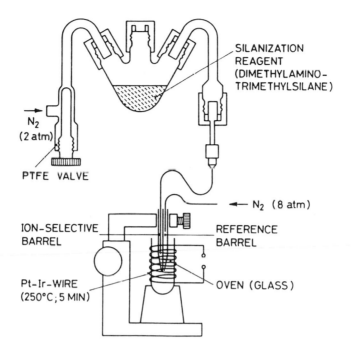

Figure 5. Apparatus for vapour-phase silanization of the ion-selective barrel of a double-barrelled microelectrode. A stream of nitrogen (2 atmospheres) carries the silane vapour through the ion-selective barrel. The reference barrel is protected against silanization by a stream of nitrogen at higher pressure (8 atmospheres). The microelectrode is placed in oven. (Modified from ref. 14 with permission.)

6.1 Front-filling micropipettes

The simplest front-filling method is to dip the silanized tip into the ion-exchanger and then backfill with the internal filling solution (*Protocol 3*). When double-barrelled microelectrodes are prepared, the reference barrel is back-filled first with an electrolyte, i.e. either saline, or 0.5 M KCl solution, or K_2SO_4/KCl solution. The choice of a reference electrolyte is limited, since it is necessary:

- to keep the changes in liquid-junction and tip potential as small as possible
- to prevent possible tissue damage by electrolyte leakage from the tip
- to achieve a low reference microelectrode resistance.

Protocol 3. Front-filling the microelectrodes

1. Dip the silanized tip of the micropipette into the ion-exchanger cocktail for approximately 10–60 seconds until an ion-exchanger column height of 100–500 µm is obtained. Wider tips fill more rapidly and thus need less time.

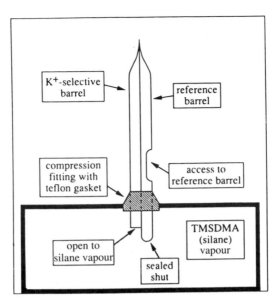

Figure 6. Schematic diagram of the theta glass silanization set-up used to fabricate double-barrelled, K$^+$-selective microelectrodes (shown in cross-section). The electrode glass is clamped in a compression fitting with a Teflon gasket, so that the back end of the electrode extends into a small chamber filled with silane vapour. The interior of the ion-selective barrel is exposed to the silane, whereas the interior of the reference barrel is not. The entire assembly is placed inside an oven during the silanization process. (From ref. 15 with permission.)

Protocol 3. *Continued*

2. For very fine tips that do not fill by capillarity, attach a syringe via plastic tubing to the untapered end of the electrode and suck the ion-exchanger in.

3. Backfill the rest of the ion-selective barrel with the internal filling solution, slowly introducing the fluid via a long syringe needle.

4. Position the electrode under a microscope and insert a fine probe towards the tip to dislodge any bubbles separating the ion-exchanger and internal filling solution. A glass fibre or a whisker (cat or rat) is ideal.

5. Remove any stubborn bubbles by heating a fine U-shaped wire in a Bunsen flame and moving it close to the outside of the glass wall near the air bubble. The localized heating dislodges the bubble.

Front-filling of ion-selective barrels is relatively easy, but usually yields ion-exchanger columns with a length of 100–500 μm. Smaller columns can also be obtained, but removal of bubbles in the narrow tip is considerably more difficult. Columns smaller than 200 μm decrease the resistance of the ISM, but are relatively difficult to achieve with front-filling procedures. The resistance can, however, be effectively decreased in coaxial microelectrodes (see *Figure 2* and ref. 9).

6.2 Back-filling micropipettes

Back-filling is also widely used today, especially when microelectrodes must have very fine tips for intracellular use. Filling requires injection of a tiny drop of ion-exchanger into the tip with a fine plastic capillary (pulled in a gas flame from plastic syringes) or by fine needles. The ion-exchanger usually spontaneously flows into the tip. If this does not happen, suction or pressure can be applied (1, 16). This type of filling results in a column several millimetres high. The injection of internal filling solution and removal of air bubbles is then easily done with a needle or a fine plastic capillary, although the probability of air bubbles between ion-exchanger and the internal filling solution decreases with column height. Frequently, air bubbles escape either spontaneously or after gentle tapping on the glass wall above them. The internal filling solutions differ according to the type of ISM, their composition is given in Section 10.

7. Microelectrode lifetime and stability

The lifetime of liquid-membrane ISMs is primarily dependent on the loss of membrane components (ion-carrier, plasticizer, solvent, additives) into the sample or storing solution. This loss affects ISM selectivity and resistance. However, in laboratory use, the lifetime is mostly dictated by mechanical defects, surface contamination (dust, bacteria, solution crystals), or by electrical leakage. Usually ISMs can be used for several days, although a lifetime of up to one month has been reported. It has also been reported that the use of quartz tubes can dramatically increase ISM lifetime (for details see ref. 6).

Instability of microelectrodes may be caused by their high impedance. Drift and noise of microelectrodes are in the range of 0.3 mV/h and 0.2 mV, respectively (17).

8. Microelectrode function and calibration

The ISM potential (*EMF*—electromotive force, mV) is described by the Nernst equation (Equation 1)

$$EMF = E_0 + s \log a_i; \tag{1}$$

where E_0 is reference potential in mV, s is Nernstian slope defined thus:

$$s = 2.303 \ RT/z_i F = 59.16/z_i \ (\text{for } 25\,°C), \tag{2}$$

where R is the gas constant $(8.314 \, \text{J K}^{-1} \text{mol}^{-1})$, T is absolute temperature (K), F is Faraday's equivalent $(9.6487 \times 10^4 \, \text{C/mol})$ and z_i is charge number of the ion I.

However, such ideal ISM behaviour does not exist, since deviation from the Nernst equation occurs at low activities. We must, therefore, consider the presence of interfering ions J present in the sample solution or tissue. The ISM potential is, therefore, far better described by the Nicolsky–Eisenman equation (18, 19):

$$EMF = E_o + E_{\Delta ljp} + s \log [a_i + \sum_{j \neq i} K_{ij}^{pot} (a_j) z_i / z_j], \tag{3}$$

where $E_{\Delta ljp}$ is the liquid-junction potential difference generated between the reference electrolyte and sample solution, z_j is the charge number of interfering ion J, and a_i and a_j are the activities of primary ion I and interfering ion J in the sample solution, respectively (in M).

The ISM exhibits a linear response at relatively high activities of ions (typically 10^{-5} to 10^{-1} M). At lower activities of the ion to be measured the calibration curve flattens. The detection limit is affected by the presence of other ions in the calibration solution. The electrodes should, therefore, be calibrated in solutions of known concentrations of the ion of interest and with approximately the same ionic composition as the sample solution or tissue compartment where the ion activity is to be measured. An example calibration curve for a K^+-ISM (Corning ion-exchanger) to be used for extracellular measurements is shown in *Figure 7*. *Figure 7* shows that when the ISM is calibrated in different KCl concentrations and with 150 mM NaCl background, the calibration curve is flattened in low K^+ activity.

The accuracy of measurements depend to a large extent on the calibration. However, small errors in the calibration procedure can be eliminated by the use of more calibration points. Moreover, in many physiological studies, the absolute values are not of primary interest since the relative changes are more important for interpretation of the results. ISMs are often calibrated for *concentration* measurements but the results are given as ion *activities*. For such a conversion, an activity coefficient has to be taken into account (for further details see refs 1–6).

9. Instrumentation

The resistance of an ISM is generally about two orders of magnitude higher than that of the reference channel $(10^8 - 10^9 \, \Omega)$. This requires a special low-noise differential amplifier with high input impedance (about $10^{14} \, \Omega$) and compensation of capacitance by negative feedback (see ref. 20). This enables accurate subtraction of the reference electrode signal (which contains other bioelectric potentials). Suitable differential amplifiers are available from WPI (World Precision Instruments, New Haven, USA). *Figure 8* shows an

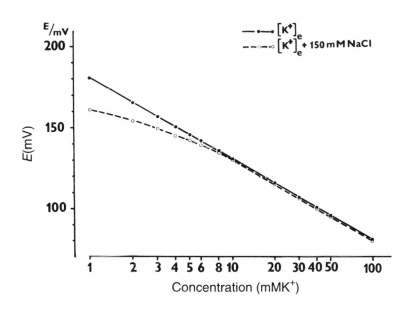

Figure 7. Calibration curve of a double-barrelled K⁺-selective microelectrode in solutions of various K⁺ concentrations. The potential changes in millivolts were plotted against concentration of K⁺ (mM) (continuous line). The dashed line represents the same microelectrode calibration in K⁺ solutions with 150 mM NaCl added to the background electrolyte. (From ref. 28 with permission.)

experimental set-up for concomitantly recording changes in extracellular K⁺ activity (aK$_e$) and E$_m$ (bioelectric potentials).

The high impedance of the ISM and its connections to the difference amplifier means that the system acts as an aerial and this can lead to problems of noise. As with other microelectrode measurements (see Chapters 1, 2, and 3) special attention must be paid to screening and earthing. It is wise to screen the preparation from external sources of interference, if necessary by using a suitably earthed Faraday cage (see Chapter 1, Section 6.3; and Chapter 3, Sections 3.1.1. and 3.1.2).

To some extent any introduced noise can be removed by filtration. It is worth bearing in mind that the signals from an ISM are essentially of very low frequency since dynamic changes occur over time scales of fractions of a second at best. Thus it is possible to low-pass the ISM signal, removing frequencies above, say, 10 Hz. This also minimizes mains interference.

Amplification of the signal may be achieved by any of several commercially available amplifiers. Again, WPI make models suitable for the purpose. Data presentation depends to some degree on the temporal response of the ISM and the biological changes being measured. There is little point using a chart recorder to record faithfully an event lasting less than a second. Conversely, an

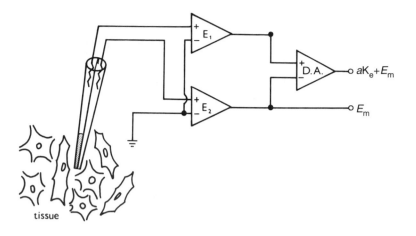

Figure 8. Diagram of experimental set-up for simultaneous recording of K^+ activity (aK_e) and E_m in a tissue. E_1 and E_2 are amplifiers. D.A.—differential amplifier for measurements with ISMs.

oscilloscope is useless for recording an entire day of experimental data. For very short duration events, suitable oscilloscopes made by Nicolet (Wisconsin, USA) (particularly the Model 310) are widely available. The Nicolet 310 also has the facility for disk-based storage of signals. Chart recorders (y–t) are mainly of use for recording slow baseline-type changes in ion concentration. Several manufacturers make adequate single- or multichannel recorders and there seems little point recommending individual manufacturers or models. The solution offering the best of both worlds is probably the microcomputer with appropriate software.

10. Available ISMs, their use and application

10.1 K⁺-selective microelectrodes

Today, we use two types of K^+-selective microelectrodes (K^+-ISM):

- based on liquid ion-exchanger Corning 477413
- based on valinomycin

Valinomycin-based ISMs have high selectivity. However, their resistance is very high (about $5 \times 10^{11} \, \Omega$) and their response time (t_{90}) is relatively slow (several seconds). ISMs using Corning 477413 have a lower resistance (about 10^9–$10^{10} \, \Omega$) and faster response time (several milliseconds). Although there is considerable interference from some cations, for example acetylcholine (ACh), tetraethylammonium (TEA⁺), or tetramethylammonium (TMA⁺), the Corning-based ISMs are most widely used for K^+ measurements in excitable

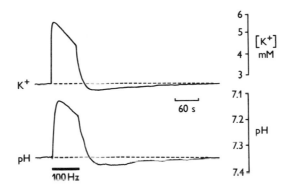

Figure 9. [K$^+$] and pH changes in dorsal horn of segment L$_4$ of rat spinal cord evoked by repetitive bipolar electrical stimulation (100 Hz, 60 s, horizontal bar) with needles inserted in plantar muscles of the ipsilateral hind paw. Recorded at a depth of 600–700 µm from dorsal spinal surface with two double-barrelled microelectrodes. The K$^+$-ISM was based on Corning 477413 and the H$^+$-ISM was based on Hydrogen Ion Ionophore II-Cocktail A (Fluka). (Modified from refs 21 and 23 with permission.)

tissues. The internal filling solution is 0.5 M KCl. The K$^+$-ISMs have recently been found to be sensitive to amiloride—an inhibitor of Na$^+$/H$^+$ exchange (23).

The upper part of *Figure 9* shows a typical example of the measurement of activity-related extracellular potassium changes in the spinal dorsal horn of the rat (21). The high density of neurons in the spinal dorsal horn and their intense activity during repetitive stimulation of peripheral nerves leads to considerable K$^+$ accumulation in the extracellular space (for review see refs 4, 22). An increase in extracellular K$^+$ concentration can also be recorded in the immediate vicinity of individual spontaneously firing neurons (*Figure 10*). The amplitude of K$^+$ changes depends on the distance of the K$^+$-ISM from the neuronal membrane and on the frequency and duration of spontaneous activity (24). If averaging is used, a K$^+$ increase can be recorded even after a *single* action potential.

10.2 H$^+$-selective microelectrodes

There are several neutral carrier ISMs for H$^+$-sensitive microelectrodes (H$^+$-ISM). Ion-selective compounds (proton ionophores) are:

- tridodecylamine (TDDA)
- 4-nonadecylpyridine (ETH 1907)

Both types of carrier have similar properties (resistance about $4 \times 10^{10}\,\Omega$, response time below 1–5 s). The ETH 1907 carrier is more suitable for low pH values (dynamic pH range 1.3–9.0) than TDDA (dynamic pH range 5.5–12.0). TDDA ionophores or cocktails must be equilibrated with CO_2, while for the

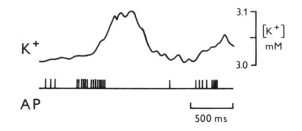

Figure 10. Top line: elevation of [K$^+$] in unstimulated reticular formation of rat associated with spontaneous bursts of cell firing. Bottom line: neuronal firing recorded by reference barrel-AP. (Modified from ref. 24 with permission.)

ETH 1907 the CO$_2$ treatment is not necessary if the additive sodium tetraphenylborate (NaTPB) is replaced by the more lipophilic potassium tetrakis (4-chlorphenyl) borate (KTpClPB). The selectivity of H$^+$-ISMs over alkali ions is high, although there is some interference of K$^+$ when using ETH 1907 intracellularly. Some H$^+$-ISMs can respond to CO$_2$, but the interference can be eliminated either by saturating the internal filling solution with CO$_2$ or using different types of H$^+$-selective cocktails (see Fluka catalogues for ionophores). The internal filling solution is composed of (in mM):

- KH$_2$PO$_4$, 40.0
- NaOH, 23.0
- NaCl, 15.0
- to pH 7.0

The lower part of *Figure 9* shows an example of extracellular pH changes in rat spinal dorsal horn evoked by stimulation of the afferent input (23). In the CNS, typical responses can be seen (either biphasic alkaline–acid or triphasic alkaline–acid–alkaline), related to various types of neuronal activity and to glial control of extracellular pH homeostasis. H$^+$-ISMs have so far been successfully used intracellularly as well as extracellularly in the nervous tissue, muscle, kidney, stomach, etc. (for review see ref. 1).

10.3 Ca^{2+}-selective microelectrodes

Currently popular Ca^{2+}-ISMs for *extracellular* measurements are based on the neutral carrier ETH 1001. (NB: Significant interference from K$^+$ and Mg^{2+} occurs during *intracellular* measurements). The resistance of Ca^{2+}-ISMs is about $2 \times 10^{10}\,\Omega$ and the electrode response time is several milliseconds. The new neutral carrier ETH 129 has a higher selectivity and is thus used for intra- as well as for extracellular measurements. As an additive, PVC is used in both types of carriers. The internal filling solution comprises 150 mM CaCl$_2$.

10.4 Na⁺-selective microelectrodes

The composition of liquid ion-exchangers for Na^+-ISMs varies according to whether they are used for intra- or extracellular measurements. Na^+-ISMs based on ETH 227 can be used for intracellular determinations, while those using ETH 157 or monensin are suitable only for extracellular measurements. ISMs based on ETH 227 and ETH 157 have limited selectivity over K^+ and Ca^{2+} and it is, therefore, necessary to calibrate them in appropriate solutions of K^+ and Ca^{2+}. The resistance of both types of microelectrodes ranges between 10^{10}–$3 \times 10^{10} \, \Omega$. The response times vary between 0.2–5.0 s. Interference to various types of inhibitors has been observed and should be checked carefully. The internal filling solution used is 0.5 M NaCl.

10.5 Cl⁻-selective microelectrodes

Cl^--ISMs can be based on:

● classical ion-exchanger Corning 477913

● a new carrier—Mn(III) porphyrin complex (5, 10, 15, 20-tetraphenyl-21H, 23H-porphin manganese (III) chloride: Mn(III)TPP.

CI⁻-ISMs based on the new carrier have better selectivity in respect to HCO_3^-, acetate, citrate, lactate, isothionate, malate, and oxalate than the Corning ion-exchanger. The resistance of ISMs based on Mn(III)TPP is high (about $7 \times 10^{10} \, \Omega$), and one should be careful of pH interference. The resistance of ISMs based on the Corning ion-exchanger is lower, their response time is shorter, but they also exhibit lower selectivity.

10.6 Other types of ion-selective microelectrodes

Other ISMs also currently available for measurements *in vivo* include Mg^{2+}-ISMs, NH_4^+-ISMs and Li^+-ISMs. These have so far been less widely used than those mentioned above.

Mg^{2+}-ISMs are based on the synthetic neutral carrier ETH 1117. Interference from K^+, Na^+ and Ca^{2+} is high and selectivity is, therefore, inadequate. The only possible way round this is to measure Na^+ and K^+ in the tissue simultaneously, and to calibrate the Mg^{2+}-ISMs in appropriate Na^+, K^+ and Ca^{2+} concentration solutions. The new improved ionophore, ETH 5214, has better selectivity and its use for intracellular measurements is recommended. However, the Mg^{2+} microelectrodes have poor selectivity towards Ca^{2+} and, therefore, they are not suitable for extracellular measurements. Their resistance is relatively high ($5 \times 10^{10} \, \Omega$), with a response time below 3 s.

NH_4^+-ISMs are based on the macrotetrolides nonactin/monactin. Their selectivity with respect to K^+ and Na^+ is poor, resistance is about $6 \times 10^{10} \, \Omega$ and response time about 1–2 s. The use of the ISMs in some tissues (for example kidney) should, therefore, be carefully evaluated.

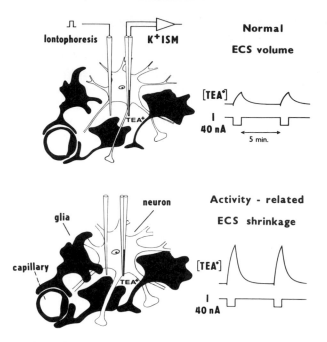

Figure 11. Scheme of experimental set-up when measuring dynamic ionic and volume changes in extracellular space (ECS). Details are given in text. [TEA+], concentration of tetraethylammonium ions; K+-ISM, K+-selective microelectrode; ECS, extracellular space.

Li+-ISMs based on the neutral carrier ETH 149 have a resistance of about 4×10^{10} Ω and are able to measure concentrations of Li+ around 1 mM in the intra- and extracellular space. Such concentrations can be found in nervous tissue during treatment of manic-depressive psychosis with Li+.

10.7 ISMs used for dynamic changes in extracellular space volume

ISMs can also successfully be used to study how substances migrate in the intracellular (ICS) and extracellular space (ECS) (see refs 25–27). Briefly, dynamic changes in the size of the ICS or ECS are studied by iontophoretic administration of ions that do not cross the cell membranes and which, therefore, remain in the ICS or ECS, their concentration being in inverse proportion to the size of the space. The potassium ion-exchanger (Corning 477317) is highly sensitive to tetraalkylammonium ions (TEA+ and TMA+) and to choline, as mentioned above, which in low concentrations are not toxic and do not cross cell membranes. They can, therefore, be used to examine changes in space volume as well as to establish absolute values, especially of the ECS volume. *Figure 11* shows, schematically, the principle of the measurement of dynamic changes in ECS volume which accompanies neuronal activity. The ions are administered by a current passing into the ECS through an iontophoretic micropipette. The K^+-ISM records ion concentration changes at an appropriate distance (40–250 μm). The changes, as well as absolute values, can be computed from the diffusion curves of the administered ion (for details see refs 25–27). This is another example of the use of ISMs based on neutral carriers.

11. Conclusions

To conclude, I would like to point out that, since the advent of their use in physiology, ISMs have provided an enormous amount of new information about the dynamics of ionic changes in living tissues (especially excitable tissues) and about the physiological significance of these changes. A few illustrative examples of their use have been presented here, but the reader can find further details in the books and reviews cited in this chapter.

References

1. Thomas, R. C. (1978). *Ion-sensitive intracellular microelectrodes*. Academic Press, London.
2. Koryta, J. (1980). *Medical and biological applications of electrochemical devices*. John Wiley, Chichester.
3. Purves, R. D. (1981). *Microelectrode methods for intracellular recording and ionophoresis, Biological techniques series*. Vol. 6. Academic Press, London, New York, Toronto, Sydney, San Francisco.
4. Syková, E., Hník, P., and Vyklický, L. (1981). *Ion-selective microelectrodes and their use in excitable tissues*. Plenum Press, New York.
5. Kessler, M., Harrison, D. K., and Hoper, J. (1985). *Ion measurements in physiology and medicine*. Springer Verlag, Berlin, Heidelberg, New York, Tokyo.
6. Ammann, D. (1986). *Ion selective microelectrodes. Principles, Design and Applications*. Springer-Verlag, Berlin, Heidelberg, New York, Tokyo.
7. Coles, J. A., Munoz, J. L., and Deyhimi, F. (1985). Surface and volume resistivity of pyrex glass used for liquid membrane ion-sensitive microelectrodes. In *Ion measurements in physiology and medicine* (eds M. Kessler, D. K. Harrison, and J. Hoper), pp. 67–73, Springer-Verlag, Berlin, Heidelberg, New York, Tokyo.
8. Thomas, R. C. (1988). *Proton passage across cell membranes*. John Wiley, Chichester.
9. Ujec, E., Keller, O., Kriz, N., Pavlik, V., and Machek, J. (1981). Double-barrel ion selective [K^+, Ca^{2+}, Cl^-] coaxial microelectrodes (ISCM) for measurements of small and rapid changes in ion activities. In *Ion-selective microelectrodes and their use in excitable tissues* (eds E. Syková, P. Hník, and L. Vyklický), pp. 41–45. Plenum Press, New York.
10. Vyskočil, F. and Kříž, N. (1972). *Pflügers Arch.*, **337**, 265.
11. Walker, J. L., Jr. (1971). *Anal. Chem.*, **43**, 89A.
12. Lux, H. D. and Neher, E. (1973). *Exp. Brain Res.*, **17**, 190.
13. Tsien, R. Y. and Rink, T. J. (1981). *J. Neurosci. Meth.*, **4**, 73–79.
14. Deyhimi, F. and Coles, J. A. (1982). *Helv. Chim. Acta*, **65**, 1752.

15. Wen, R. and Oakley, B., II (1990). *J. Neurosci. Meth.*, **31**, 207.
16. Lanter, F., Erne, D., Ammann, D., and Simon, W. (1980). *Anal. Chem.*, **52**, 2400.
17. Bührer, T., Gehring, P., and Simon, W. (1988). *Anal. Sci.*, **4**, 547.
18. Nicolsky, B. P. (1937). *Zh. Fis. Khim.*, **10**, 495.
19. Eisenman, G. (1967). *Glass electrodes for hydrogen and other cations: principles and practice.* M. Dekker, New York.
20. Ujec, E. (1988). *Physiol. Bohemoslov.*, **37**, 87.
21. Svoboda, J., Moton, V., Hájek, I., and Syková, E. (1988). *Brain Res.*, **458**, 97.
22. Syková, E. (1983). *Prog. Biophys. Mol. Biol.*, **42**, 135.
23. Syková, E. and Svoboda, J. (1990). *Brain. Res.*, **512**, 181.
24. Syková, E., Rothenberg, S., and Krekule, I. (1974). *Brain Res.*, **79**, 333.
25. Nicholson, C., Phillips, J. M., and Gardner-Medwin, A. R. (1979). *Brain Res.*, **169**, 580.
26. Nicholson, C. and Phillips, J. M. (1981). *J. Physiol.*, **321**, 225.
27. Nicholson, C. and Rice, M. E. (1988). Use of ion-selective microelectrodes and voltammetric microsensors to study brain cell microenvironment. In *Neuromethods: the neuronal microenvironment* (eds A. A. Boulton, G. B. Baker, and W. Walz), pp. 247–361. The Humana Press, New York.
28. Kříž, N., Syková, E., and Vyklický, L. (1975). *J. Physiol.*, **249**, 167.

A1

Suppliers of specialist items

The list given below covers manufacturers and suppliers of *specialist* apparatus.

Aldrich Chemical Co., New Road, Gillingham, Dorset SP8 4JL, UK.

Amicon Division, W. R. Grace & Co., 72 Cherry Hill Drive, Beverly, MA 0915, USA.

Analog Devices Inc., P.O. Box 280, Norwood, MA 02062, USA.

AVCO Textron Special Materials, 2 Industrial Avenue, Lowell, MA 01851, USA.

Axon Instruments, 1101 Chess Drive, Foster City, CA 94404, USA.

Bak Electronics Inc., 13220-N2 Wisteria Drive, Germantown, MD 20874, USA.

Barr & Stroud Ltd., Caxton Street, Anniesland, Glasgow, G13 1HZ, UK.

Beckman Inst Inc., 2500 Harbor Blvd, Fullerton, CA 92634, USA.

Becton Dickinson & Co., 299 Webro Road, Parsippany, NJ 07054, USA.

Bioanalytical Systems (BAS) Inc., 2701 Kent Avenue, West Lafayette, IN 47906, USA.

Bio-Logic, 4 Rue Docteur Pascal, 38130 Echirolles, France.

Biotech Instruments Ltd., 183 Camford Way, Luton, LU3 3AN, UK.

Burleigh Instruments, Burleigh Park, Fishers, NY 14453, USA.

Burr–Brown Corporation, International Airport Industrial Park, P.O. Box 11400, Tucson, AZ 85734, USA.

Calbiochem, Novabiochem (UK) Ltd, 3 Heathcoat Building, Highfields Science Park, University Boulevard, Nottingham NG7 2QJ, UK.

Calbiochem Corp., P.O. Box 12087, San Diego, CA 92112-4180, USA.

Campden Instruments, 186 Campden Hill Road, London W8 7TH, UK.

Carnegie Medicin, Roslagsvägen 101, S-104 05 Stockholm, Sweden.

CED Ltd., Science Park, Milton Road, Cambridge CB4 4FE, UK.

C. G. Processing Inc., Box 133, Rockland, DE 19732, USA.

Chrompack International B. V., P.O. Box 8033, NL-4330 EA Middelburg, The Netherlands.

Ciba-Corning Diagnostics Ltd., Colchester Road, Halstead, Essex CO9 2DX, UK.

Clark Electromedical, P.O. Box 8, Pangbourne, Reading RG8 7HU, UK.

Cohu Inc., Electronics Division, 5755 Kearny Villa Road, San Diego, CA 92123, USA.

Cole Parmer (CP Instrument Co. Ltd.,), P.O. Box 22, Bishops Stortford, Herts CM23 3DX, UK.

Cole Parmer Instrument Co., 7425 Nth Oak Park Ave, Chicago, IL 60648, USA.

Courtaulds Ltd., Carbon fibres Unit, P.O. Box 16, Coventry CV6 5AE, UK.

Dagan Corporation, 2655 Park Avenue, Minneapolis, MN 55407, USA.

Dow Corning Corp., Dow Corning Centre, Box 0994, Midland, MI 48686-0994, USA.

Du Pont de Nemours International S.A., CH–1211 Geneva 24, Switzerland.

Ealing Co., 22 Pleasant Street, South Natick, MA 01760, USA.

Eastman Kodak Co., 343 State Street, Rochester, NY 14650, USA.

EG & G Instruments, Sorbus House, Mulberry Business Park, Wokingham, Berks RG11 2GY, UK.

Enka America Inc., P.O. Box 2659, 1 North Park Square, Asheville, NC 28802, USA.

Ensman Instrumentation, 4151 Broadway, Bloomington, IN 47403, USA.

Fine Science Tools, 323B Vintage Park Drive, Foster City, CA 94404, USA.

Fluka Chemical Corp., 980 South Second Street, Ronkonkama, NY 11779, USA.

Fluka Chemicals Ltd., Peakdale Road, Glossop, Derbyshire SK13 9XE, UK.

Frequency Devices, 25 Locust Street, Haverhill, MA 01830, USA.

Goodfellows Metals Ltd., Cambridge Science Park, Cambridge CB4 4DJ, UK.

Gould Electronics Ltd, Roebuck Road, Hainault, Ilford IG6 3UE, UK.

Gould Inc., Recording Systems Div., 3631 Perkins Avenue, Cleveland, OH 44114, USA.

Frederick Haer, Brunswick, ME 04011, USA.

Hamamatsu Photonic Systems Corp., 360 Foothill Road, P.O. Box 6910, Bridgewater, NJ 08807-8450, USA.

Harvard Apparatus Inc., 22 Pleasant Street, Natick, MA 01760, USA.

Harvard Apparatus Ltd., Fircroft Way, Edenbridge, Kent TN8 6HE, UK.

Hewlett Packard Co., Analytical Products Group, 3000 Hanover Street, Palo Alto, CA 94304, USA.

A. R. Horwell Ltd., 73 Maygrove Road, West Hampstead, London NW6 2BP, UK.

Hospal Ltd., 8A Consul Road, Dribune Industrial Estate, Rugby, UK.

Imaging & Sensing Technology Corp., Westinghouse Centre, Horseheads, NY 14845, USA.

Indec Systems Inc., 1283A Mt View-Alviso Road, Sunnyvale, CA 94089, USA.

Intracel Ltd., Unit 4, Station Road, Shepreth, Royston, Herts SG8 6PZ, UK.

Jencons Scientific Ltd., Cherrycourt Way Industrial Estate, Stanbridge Road, Leighton Buzzard, Beds LU7 8UA, UK.

Aesar Johnson Matthey Inc., Eagles Landing, P.O. Box 1087, NH 03874, USA.

Johnson Matthey Ltd., Orchard Road, Royston, Herts SG8 5HE, UK.

Joyce–Loebl Ltd, Dukesway, Team Valley, Gateshead, Tyne and Wear, NE11 0PZ, UK.

David Kopf, 7324 Elmo Street, P.O. Box 636, Tujunga, CA 91042-0636, USA.

Lamp Metals Ltd., Fourth Avenue, Team Trading Estate, Gateshead NE11 0TY, UK.

Leitz Gmbh, D-6330, Wetzlar, Germany.

Lipshaw Manufacturing (Shandon Lipshaw Inc.), 171 Industry Drive, Pittsburgh, PA 15275, USA.

List Electronic, Pfungerstaedter Strasse 18–20, D-6100 Darmstadt, Germany.

Mecanex, Vuarpilliere 29, CH-1260, Nyon, Switzerland.

Micro Instruments Ltd., 18 Hanborough Park, Long Hanborough, Oxford OX7 2LH, UK.

Molecular Probes, 4849 Pitchford Ave., Eugene, OR 97402, USA.

Narashige Co., 9 28 Kasuya 4 Chome, Setagaya-ku, Tokyo 157, Japan.

Narashige USA, 1 Plaza Road, Greenvale, NY 11548, USA.

Newcastle Photometric Systems, 18 Windsor Terrace, Jesmond, Newcastle upon Tyne NE2 4HE, UK.

Newport, 18235 Mt Baldy Circle, Fountain Valley, CA 92708, USA.

Newport Ltd., 12 Spa Industrial Park, Royal Tunbridge Wells TN2 3EP, UK.

Nicolet Instrument Corp., 5225 Verona Road, Madison, WI 53711-0451, USA.

Nicolet Instruments Ltd., Budbrooke Road, Warwick CV34 5XH, UK.

Nikon Europe B.V., Instrument Dept., P.O. Box 222, NL-1170 AE Badhoevedorp, The Netherlands.

Nikon Inc., Instrument Group, 1300 Walt Whitman Road, Melville, NY 11747-3064, USA.

Nippon Kankoh-Shikiso Kenykyusho Co. Ltd., Okayama, Japan.

Oriel Corporation, 250 Long Beach Blvd, Stratford, CT 06497, USA.

PD Systems, Estate House, Pool Close, West Molesey, Surrey KT8 0RN, UK.

Pharmacia LKB Biotechnology Inc., 800 Centennial Avenue, P.O. Box 1327, Piscataway, NJ 08855-1327, USA.

Pharmacia Ltd., Midsummer Boulevard, Central Milton Keynes MK0 3HP, UK.

Photometrics Ltd., 3440 East Britannia Drive, Tucson, AZ 85706, USA.

Racal Recorders Inc., 5 Research Place, Rockville, MD 20850, USA.

Racal Recorders Ltd., Hardley Industrial Estate, Hythe, Southampton SO4 6ZH, UK.

Radio Spares (RS) Components Ltd., P.O. Box 99, Corby, Northants NN17 9RS, UK.

Rainin Inst. Co. Inc., 1715 64th Street, Emeryville, CA 94608, USA.

Razel Scientific Instruments Inc., 100 Research Drive, Stamford, CT 06906, USA.

Roth Industrial Instruments Ltd., Alpha House, Alexandra Road, Farnborough, Hants GU14 6BU, UK.

Schott Glaswerke, Chemical Division, Laboratory Product Group, Hattenbergstrasse 10, Postfach 2480, D-6500 Mainz, Germany.

Scientific Glass Engineering Ltd., Potters Lane, Kiln Farm, Milton Keynes MK11 3LA, UK.

SPEX, Glen Spectra Ltd., 2–4 Wigton Gardens, Stanmore, Middx HA7 1BG, UK.

Sutter Instrument Co., 40 Leveroni Ct, Novato, CA 94949, USA.

Technical Manufacturing Corp. (TMC), 15 Centennial Drive, Peabody, MA 01960, USA.

Tektronix Inc., P.O. Box 500, Beaverton OR 97077, USA.

Vincent Associates, 1255 University Avenue, Rochester, NY 12607, USA.

Wentworth Co., Sunderland Road, Sandy, Beds SG19 1RB, UK.

Windsor Scientific, 854 Plymouth Road, Slough Trading Estate, Slough, Berks SL1 4LP, UK.

WP Instruments, 175 Sarasota Centre Blvd, Sarasota, FL 34240, USA.

WP Instruments Ltd., Undercliffe House, Studio 1, Rock-a-nore, Hastings, East Sussex TN34 3DW, UK.

Carl Zeiss Inc., 1 Zeiss Drive, Thornwood, NY 10594, USA.

Index